Theological Neuroethics

T&T Clark Enquiries in Theological Ethics

Series editors

Brian Brock

Susan F. Parsons

By the same author:

Study Guide to Christian Ethics (London: SCM Press, 2006).

Selfish Genes and Christian Ethics: Theological and Ethical Reflections on Evolutionary Biology (London: SCM Press, 2007).

Respecting Life: Theology and Bioethics (London: SCM Press, 2011).

Flourishing: Health, Disease and Bioethics in Theological Perspective (Grand Rapids, MI: Eerdmans, 2013).

Theological Neuroethics

Christian Ethics Meets the Science of the Human Brain

Neil Messer

Bloomsbury T&T Clark
An imprint of Bloomsbury Publishing Plc

B L O O M S B U R Y
LONDON • OXFORD • NEW YORK • NEW DELHI • SYDNEY

Bloomsbury T&T Clark
An imprint of Bloomsbury Publishing Plc

Imprint previously known as T&T Clark

50 Bedford Square
London
WC1B 3DP
UK

1385 Broadway
New York
NY 10018
USA

www.bloomsbury.com

BLOOMSBURY, T&T CLARK and the Diana logo are trademarks of Bloomsbury Publishing Plc

First published 2017

British Library Cataloguing-in-Publication Data
A catalogue record for this book is available from the British Library.

ISBN: HB: 978-0-5676-7139-4
ePDF: 978-0-5676-7141-7
ePub: 978-0-5676-7140-0

Library of Congress Cataloging-in-Publication Data
LC record available at http://lccn.loc.gov/2015040645

Series: T&T Clark Enquiries in Theological Ethics

Typeset by Integra Software Services Pvt. Ltd.

In memoriam
Kenneth Brian Wilson
10th April 1937–12th January 2017

Contents

Acknowledgements

It is a pleasure to acknowledge the help, advice and friendship from many quarters that have supported me in this work. Most of the book was written during six months' research leave from January to June 2016, granted by the University of Winchester on the completion of a six-year term as Head of the Department of Theology, Religion and Philosophy. I am very grateful to the University for making this possible. I am also acutely conscious of the extra burdens that several colleagues had to bear while I was away: heartfelt thanks to them in particular.

During that period of research leave, I spent an enjoyable and productive five weeks as a Guest Professor in the Center for Theology, Science, and Human Flourishing at the University of Notre Dame, Indiana, USA. I am most grateful to the University of Notre Dame for this appointment. Sincere thanks to Center director Celia Deane-Drummond for making my visit possible and for her and her family's welcome and hospitality, as well as her friendship and interest in my work over many years. I enjoyed good conversations with Celia and several of her colleagues while at Notre Dame, and I am grateful to all those who gave me their time and showed an interest in this project. My thanks also to Becky Artinian-Kaiser and Katie Zakas Rutledge for facilitating the administrative arrangements for my visit, and to Becky and her husband Josh for their hospitality and friendship during my stay.

When I was first planning to work on neuroethics, a seminar paper at the Ian Ramsey Centre, University of Oxford, and my professorial inaugural lecture at the University of Winchester helped me get my thoughts into shape. Subsequently, parts of the book have been delivered as lectures, seminars and conference papers at the Faraday Institute, Cambridge, the University of Nottingham, St Mary's University, London, the University of Winchester, the Association of Teachers of Moral Theology, the Science and Religion Forum, the Society of Christian Ethics, the Society for the Study of Christian Ethics, and the Worthing Theological Society. I am grateful for these invitations and opportunities, and for all that I learned from the discussions that they made possible.

Some material from Chapters 2 and 3 appears in shortened form as: 'What should Christian Theology (not) Learn from Science? The Case of the Human Brain', in Gillian Straine (ed.), *The Limits of Science* (Newcastle upon Tyne: Cambridge Scholars Publishing, 2017).

Chapters 3 and 4, respectively, are much expanded and developed versions of material first published in two papers:

'Cognitive Science, Moral Reasoning, and the Theological Suspicion of Ethics', *Journal of the Society of Christian Ethics*, 36.1 (2016), pp. 51–68.
'Determinism, Freedom and Sin: Reformed Theological Resources for a Conversation with Neuroscience and Philosophy', DOI: 10.1177/0953946815570591. The final, definitive version of this article is published in *Studies in Christian Ethics*, 28.2 (May 2015), pp. 163–74, published by Sage Publishing, all rights reserved.

Permission to reuse copyright material is gratefully acknowledged.

I thank Celia Deane-Drummond and Roy Weller for reading and commenting on various draft chapters. I am grateful to Susan Parsons and Brian Brock, Series Editors of T&T Clark's *Enquiries in Theological Ethics*, for enthusiastically supporting my proposal for this contribution to the series, and for their comments, advice and criticism, which have undoubtedly improved the final product. It goes without saying that I bear sole responsibility for all its remaining flaws and deficiencies. Thanks to Anna Turton and her colleagues at Bloomsbury T&T Clark for all they have done to enable the proposal to become a published book.

The dedication honours the memory of my former colleague Kenneth Wilson, who died in January 2017, not long before I finished working on this book. Kenneth and I were colleagues at the Queen's Foundation, Birmingham, from 1998 to 2001, at the outset of my academic career in theology. His generous encouragement and advice did much to give me the confidence to follow an academic theological path. He particularly encouraged me to bring my previous research background in the biosciences into dialogue with my research interests in theological ethics: in short, just the kind of work I have attempted in this book as well as many of my previous publications. Kenneth often described theological work as having good conversations, and his infectious enthusiasm for such conversations has played a big part in shaping my own sense of what I am about as a scholar and teacher. May he rest in peace and rise in glory.

And finally, heartfelt thanks to Janet, Fiona and Rowan: most of all, simply for being there; more particularly for the love, understanding and support they have shown during the writing of this book; not least, for bearing with the various tiresome effects of having an academic in the house.

Neil Messer
Winchester
Lent 2017

1

Introduction

Why brain science matters to Christian ethics – and vice versa

In 1990, US President George H. W. Bush declared the 1990s the 'Decade of the Brain'.[1] The years since the president's proclamation have seen remarkable developments in neuroscience, advancing the understanding of many aspects of brain function and their impacts on human life.[2] Some of these advances in understanding, as we shall see, raise intriguing and sometimes troubling questions about human nature and identity. Others bring with them increasing prospects for modifying the functions of the brain, either to treat diseases and disorders or in the hope of enhancing human functioning. These possibilities, of course, raise intriguing and troubling questions of their own, about the proper scope and limits of such interventions.

Along with the rapid development of neuroscience and neurotechnology has come an explosive growth of public interest in the science of the brain. Popular books, television documentaries, media reports and online coverage of neuroscience have proliferated in the decades since the 'Decade of the Brain'.[3] It is not difficult to see why. Whatever view one takes of the relationship between brain, mind and soul, it is evident that our brains play a distinctive and essential part in more or less every aspect of our lives as human persons, from the regulation of the basic functions that keep us alive to emotions and desires,

[1] George Bush, Presidential Proclamation 6158 (17 July 1990), online at http://www.loc.gov/loc/brain/proclaim.html (accessed 13 March 2017).

[2] For a survey of some of these, see Nora D. Volkow et al., 'A Decade after the Decade of the Brain', *Cerebrum* (26 February 2010), online at http://dana.org/Cerebrum/2010/A_Decade_after_The_Decade_of_the_Brain__Compilation/ (accessed 24 February 2017).

[3] One high-profile recent example is David Eagleman, *The Brain with David Eagleman* (BBC4, 6 parts, 21 January–26 February 2016), and the associated book: David Eagleman, *The Brain: The Story of You* (Edinburgh: Canongate, 2015).

thinking, willing and acting, understanding and beliefs about the world, self-understanding and personal identity, personal and social relationships, and even our experience and practice of faith. A brain disorder may affect someone's personality, relationships and sense of identity so profoundly that those who know her wonder whether she is still the same person. Mental illnesses are increasingly conceptualized as disorders of neural pathways or circuits in the brain.[4] Some brain injuries and diseases can either temporarily or permanently remove most or all of their victims' capacity for consciousness, awareness and interaction with their surroundings; in cases where that loss is permanent, it has become almost commonplace to ask whether there is really a *person* there anymore. In short, understanding our brains seems to promise a great deal by way of understanding *ourselves*. No wonder so many people find neuroscience so interesting.

From this burgeoning growth of neuroscience has emerged the discipline of neuroethics. The word may still be unfamiliar in some quarters,[5] but since its beginnings in the early 2000s the field of neuroethics has become a sizeable and growing academic industry, complete with international societies, journals, monographs, textbooks and so forth. Of course, some of the topics discussed in these books and journals (including some that will be discussed in this book) have longer histories than neuroethics itself, having long been on the agenda of related fields such as bioethics. But there are grounds for thinking there is something distinctive about the questions raised by neuroscience and its clinical applications. In an article which was more or less the founding charter of neuroethics, Adina Roskies put it like this: '[t]he intimate connection between our brains and our behaviors, as well as the peculiar relationship between our brains and our selves, generate distinctive questions that beg for the interplay between ethical and neuroscientific thinking'.[6]

Roskies divided this emerging field into the 'ethics of neuroscience' and the 'neuroscience of ethics'.[7] In the 'ethics of neuroscience', she included questions

4 Volkow et al., 'A Decade after the Decade of the Brain'.

5 While writing this book, I had several conversations with friends and colleagues who began to look puzzled when I said I was working on 'neuroethics'. Some mis-heard me and thought I was writing about 'Euroethics'; since most of the book was written in the year of the UK's Brexit referendum, this had the potential to cause all kinds of misunderstanding.

6 Adina Roskies, 'Neuroethics for the New Millennium', *Neuron* 35 (2002), pp. 21–23.

7 Ibid.

about the ethical conduct of neuroscientific research and the use of its findings, but also the ethical implications of neuroscientific findings. For example, should findings about neural correlates of antisocial or criminal behaviour have any bearing on our ethical conclusions about criminal justice and punishment, and if so, what? The 'neuroscience of ethics' referred to such things as the scientific study of moral reasoning, judgement and action. However, Roskies also remarked that the 'most intriguing' aspects of neuroethics might be how the two parts of the field interacted with each other.[8] Indeed, this book will be concerned with questions in both parts of the field as mapped by Roskies, but also with interactions between the two.

Talk of neuroscience and its ethical implications, of course, is hardly immune from muddled and distorted thinking, hype, sensationalism and other such pathologies. Some critics complain of 'neuromania', the fashion for attaching a 'neuro' prefix to anything from marketing to theology and assuming that this will open up new vistas of understanding.[9] Others criticize the 'neuroessentialism' – the assumption that 'our brains define who we are'[10] – often found in both academic literature and popular discourse about neuroscience.[11]

Certainly, it is easy for exaggerated or sensationalized accounts to distort public perceptions of the scope, power and implications of neuroscience. And since neuroscientific studies of complex mental phenomena are inevitably theory-laden at every stage from conceptualization through design to analysis and interpretation of the results, there is a real risk that naïve or muddled philosophical – or indeed theological – assumptions may result in ill-conceived and distorted research studies. Indeed, some areas of research discussed in this book are prone to such problems – as critics often maintain, and as I shall point out from time to time in the following chapters. But such problems should

[8] Ibid., p. 21.

[9] Paulo Legrenzi and Carlo Umtilà, *Neuromania: On the Limits of Brain Science* (Frances Anderson (trans.); Oxford: Oxford University Press, 2011).

[10] Roskies, 'Neuroethics for the New Millennium', p. 22.

[11] For an example from popular literature, one need look no further than the subtitle of the book by David Eagleman cited above: *The Brain: The Story of You*. In similar vein, the final sentence of his Introduction reads: 'In the infinitely dense tangle of billions of brain cells and their trillions of connections, I hope you'll be able to squint and make out something that you might not have expected to see there. You' (ibid., p. 2). For a vigorous critique of neuroessentialist assumptions in the ethical literature, see Harris Wiseman, *The Myth of the Moral Brain: The Limits of Moral Enhancement* (Cambridge, MA: MIT Press, 2016), ch. 1.

not distract us from the achievements and future promise of neuroscience.[12] It does seem to raise real, fascinating and perhaps disturbing questions about our self-understanding, our possibilities for living and acting in the world and the possibilities and proper limits of technological interventions based on neuroscientific understanding.

All this should be of great interest to Christian theology and ethics. Close to the heart of Christian theology are questions about what it means to be a human creature before God, in relation to others and in the world we inhabit. An area of science offering claims and insights about human nature, identity, personality and relationships should, one would think, attract the interest of Christian theologians concerned about such things. Moreover, at times, neuroscience turns the spotlight directly onto Christian faith and practice, with studies of the brain activity, processes and mechanisms associated with religious experience and belief. What significance – if any at all – do such studies have for the truth or validity of the beliefs and practices they are concerned with?

There has indeed been notable (though also sometimes problematic) theological work on some of these questions, particularly in the science and theology literature. Various authors have attempted different kinds of dialogue or apologetic engagement with neuroscience.[13] Not surprisingly, one specific topic that has attracted significant attention is the implications of neuroscience for a Christian understanding of the soul, which will prove relevant to one of

[12] There is a parallel with slightly earlier developments in human genetics. Since the late 1970s, that field too has seen remarkable developments in its methods and techniques, the knowledge and understanding gained using those methods, and possibilities for technological intervention and manipulation; and as with neuroscience, the science and the technological possibilities continue to grow apace. Along with these scientific and technological developments has come a good deal of hype in both popular and academic discourse: we were often assured in the 1990s, for example, that the sequencing of the human genome would reveal what it really means to be human. With the passage of time, much of this hype has been deflated; what remains, however, is a clearer view of the real recent achievements and future prospects of human genetics, and the real – and sometimes perplexing – ethical questions these achievements and prospects pose.

[13] For example, Warren S. Brown and Brad D. Strawn, *The Physical Nature of Christian Life: Neuroscience, Psychology and the Church* (Cambridge: Cambridge University Press, 2012); Oliver Davies, *Theology of Transformation: Faith, Freedom and the Christian Act* (Oxford: Oxford University Press, 2013); Markus Mühling, *Resonances: Neurobiology, Evolution and Theology. Evolutionary Niche Construction, the Ecological Brain and Relational-Narrative Theology* (Göttingen: Vandenhoeck and Ruprecht, 2014); Andrew B. Newberg, *Principles of Neurotheology* (Farnham: Ashgate, 2010); Gregory R. Peterson, *Minding God: Theology and the Cognitive Sciences* (Minneapolis, MN: Fortress, 2003); Robert John Russell et al. (eds.), *Neuroscience and the Person: Scientific Perspectives on Divine Action* (Vatican City: Vatican Observatory Publications/Berkeley, CA: Center for Theology and the Natural Sciences, 1999).

the areas explored later in this book.[14] But surprisingly little of the theological engagement with neuroscience has addressed *neuroethical* topics in any depth. This is not to say, of course, that such work has been completely absent. For example, authors such as Nancey Murphy have explored freewill and responsibility in light of neuroscience.[15] Theologians interested in bioethics have engaged with practical topics like the treatment of patients in vegetative state[16] and the ethics of human enhancement and transhumanism[17] – though theological discussion of human enhancement has tended to focus more on genetic than neural enhancements.[18] Many of these discussions, however, have been confined to the more familiar territory of Roskies's 'ethics of neuroscience', while fewer have dealt with the 'neuroscience of ethics' or the intersections between the two.[19] Authors who set out to engage theologically with the field of neuroethics as a whole are not very easy to find.

Why might this be? No doubt there are various reasons. Some theological traditions and approaches might be sceptical in principle that neuroscience and theological ethics have much to say to each other. Karl Barth – who will be one of my own main theological conversation partners in this book – was famously dismissive of the relevance of the natural sciences to Christian doctrine,[20] and

[14] For example, Warren S. Brown, Nancey Murphy and H. Newton Moloney (eds.), *Whatever Happened to the Soul? Scientific and Theological Portraits of Human Nature* (Minneapolis, MN: Fortress, 1998); Joel B. Green (ed.), *What about the Soul? Neuroscience and Christian Anthropology* (Nashville, TN: Abingdon, 2004); Malcolm Jeeves (ed.), *From Cells to Souls – and Beyond: Changing Portraits of Human Nature* (Grand Rapids, MI: Eerdmans, 2004); Keith Ward, *In Defence of the Soul* (Oxford: Oneworld, 1998).

[15] For example, Nancey Murphy and Warren S. Brown, *Did My Neurons Make Me Do It? Philosophical and Neurobiological Perspectives on Moral Responsibility and Free Will* (Oxford: Oxford University Press, 2007).

[16] For example, Jeffrey P. Bishop and Elliott Louis Bedford (eds.), Theme Issue, 'Medically Assisted Nutrition and Hydration: The Vegetative State and Beyond', *Christian Bioethics* 17.2 (2011); Pia Matthews, *Pope John Paul II and the Apparently 'Non-Acting' Person* (Leominster: Gracewing, 2013).

[17] For example, Celia Deane-Drummond and Peter Manley Scott (eds.), *Future Perfect? God, Medicine and Human Identity* (London: T&T Clark, 2006); Brent Waters, *From Human to Posthuman: Christian Theology and Technology in a Postmodern World* (Farnham: Ashgate, 2006); Waters, *Christian Moral Theology in the Emerging Technoculture: From Posthuman Back to Human* (Farnham: Ashgate, 2014).

[18] For example, Corinna Delkeskamp-Hayes and Neil Messer (eds.), Theme Issue, 'Theological Anthropology and the Ethics of Human Germline Genetic Modification', *Christian Bioethics* 18.2 (2012).

[19] Those that do include John Perry, 'Jesus and Hume among the Neuroscientists: Haidt, Greene, and the Unwitting Return of Moral Sense Theory', *Journal of the Society of Christian Ethics* 36.1 (2016), pp. 69–85; Sally K. Severino, *Behold Our Moral Body: Psychiatry, Duns Scotus, and Neuroscience* (London: Versita, 2013); James A. Van Slyke et al. (eds.), *Theology and the Science of Moral Action: Virtue Ethics, Exemplarity, and Cognitive Neuroscience* (New York, NY: Routledge, 2013).

[20] See Karl Barth, *Church Dogmatics* (G. W. Bromiley and T. F. Torrance (ed.), 13 vols., Edinburgh: T&T Clark, 1956–75), vol. 3.1, pp. ix–x.

some of his readers continue to adopt a similar stance in relation to Christian ethics. Some Christian ethicists might be repelled by the neuroessentialism that, as I have already noted, is easy to find in neuroethics. They might conclude (with some justice) that this is an academic discourse set up to exclude theological perspectives and insights. A related reason could be impatience with bad empirically based arguments about normative questions.[21] I agree that neuroethical literature too often displays these problematic features, but I shall suggest below that this is not a good reason to avoid it.

Neuroethics, for its part, largely ignores theology. As it has developed since the early 2000s, it is a remarkably secular field: though it is highly interdisciplinary, it is quite rare for theology to be one of the disciplines that feature in the conversation. A text like the *Oxford Handbook of Neuroethics* vividly illustrates this:[22] the list of almost 100 contributors includes many neuroscientists, psychologists, clinicians, philosophers, bioethicists and legal scholars, but no one who is identified as a theologian.[23] No chapter titles explicitly signal religious or theological themes, and neither 'religion' nor 'theology' appears in the index. My point is not to single out this work, or its contributors, but to suggest that it is symptomatic of the field's secularity that the editors of a major authoritative text apparently did not see any need to include theology among the disciplines represented in the book.[24]

[21] This objection is discussed by Perry, 'Jesus and Hume among the Neuroscientists', pp. 77–78, in dialogue with Oliver O'Donovan, *Self, World, and Time: Ethics as Theology* (Grand Rapids, MI: Eerdmans, 2013).

[22] Judy Illes and Barbara J. Sahakian (eds.), *The Oxford Handbook of Neuroethics* (Oxford: Oxford University Press, 2011).

[23] This is not to suggest none of the contributors is aware of, or sympathetic to, theological themes: one, the neuroimmunologist Alasdair Coles, is identified in his author biography as also a Church of England minister, while another, George Khushf, is a philosopher who has published on theological themes; ibid., pp. xxiv, xxviii. A similar lack of theological contributions is evident in the multi-part neuroethics bibliography complied by James Giordano and his colleagues; among around 2,000 books, chapters and journal articles, only a handful of titles or other features indicate that they include theological perspectives (although of course some might include such perspectives without signalling it in their titles): Liana Buniak, Martina Darragh and James Giordano, 'A Four-Part Working Bibliography of Neuroethics: Part 1: Overview and Reviews – Defining and Describing the Field and Its Practices', *Philosophy, Ethics, and Humanities in Medicine* 9.9 (2014), DOI: 10.1186/1747-5341-9-9; Martina Darragh, Liana Buniak and James Giordano, 'A Four-Part Working Bibliography of Neuroethics: Part 2 – Neuroscientific Studies of Morality and Ethics', *Philosophy, Ethics, and Humanities in Medicine* 10.2 (2015), DOI: 10.1186/s13010-015-0022-0; Amanda Martin et al., 'A Four-Part Working Bibliography of Neuroethics: Part 3 – "second tradition neuroethics" – Ethical Issues in Neuroscience', *Philosophy, Ethics, and Humanities in Medicine* 11.7 (2016), DOI: 10.1186/s13010-016-0037-1.

[24] For relatively rare exceptions, see (e.g.) Stephanie J. Bird, Michael Kalichman and Dena Plemmons (eds.), Theme Issue, 'Multidisciplinary Perspectives on Neuroethics', *Science and Engineering Ethics* 18.3 (2012), and Wiseman, *The Myth of the Moral Brain*.

Again there could be various reasons for this neglect, but it seems likely that neuroessentialist assumptions play a part. If it is assumed that 'our brains define who we are' (as Roskies puts it), then it seems natural to conclude that sooner or later the scientific study of our brains will tell us everything there is to be learned about who we are, and that religious or theological accounts of what it is to be human are beside the point. Some neuroscientists think that neuroscience denies the existence of the soul and thereby discredits religious views of human nature.[25] And perhaps, some neuroethicists imagine that neuroscience discredits theology – making it irrelevant to neuroethics – by offering scientific explanations for the existence of religious belief. Needless to say, I regard claims and assumptions like these as very poor reasons to think that Christian theology is either discredited or irrelevant to neuroethics. The next chapter of this book will be devoted in part to addressing such claims head-on, in order to clear the ground for theological engagement with neuroethics itself.

Whatever its causes, the mutual neglect of neuroethics and Christian ethics is detrimental to both. Neuroethicists do not have good reasons to dismiss Christian theology as either discredited or irrational. Therefore, they cannot simply assume that Christian traditions of moral reasoning are irrelevant to neuroethical enquiry. Nor would they be justified in thinking that theological approaches are only of interest to religious believers and communities, and are unfit for 'public' ethical discourse.[26] They might even find that theological approaches have interesting and helpful things to offer. In the following chapters, I shall give various examples of neuroethical discussions framed in unhelpful or distorting ways, and argue that by reframing such discussions, theological ethics can offer more satisfying solutions to the problems they are concerned with.

Conversely, theologians would be ill-advised to dismiss neuroscience as irrelevant to Christian ethics. If Christian ethics is concerned with what it means to live well as an embodied human creature before God, in company with others, in the physical world that we inhabit, it seems unwise simply to

[25] See, e.g., Francis Crick, *The Astonishing Hypothesis: The Scientific Search for the Soul* (New York, NY: Charles Scribner's Sons, 1994). As we shall see in Chapter 5, however, matters are a little more complicated than Crick imagined.

[26] For one among many rebuttals of such a view, see Nigel Biggar, *Behaving in Public: How to Do Christian Ethics* (Grand Rapids, MI: Eerdmans, 2011).

ignore areas of scientific enquiry that make claims about the conditions of such physically embodied moral lives. Of course, as I have already acknowledged, neuroethical discourse includes its share of tendentious assumptions, bad arguments and suchlike faults. But this is surely a reason for Christians to join in the argument rather than eschewing it. Indeed, as we shall see, neuroethical arguments underpin matters of vital practical concern in health care and other fields, and neuroethicists address not only academic debates but also questions of professional practice and public policy. If Christian ethicists do not wish to see these questions shaped by careless or tendentious arguments, it behoves them to contribute better ones, in any way that their theological vantage points equip them to. It must also be acknowledged, though, that there are better and worse ways of engaging theologically with neuroscience, and some of the ways to be found in the literature seem to me to be more or less false trails. One purpose of this book is to identify what I see as more satisfactory ways of going about the task.

If Christian ethicists do attempt this kind of theological engagement, they might get some unexpected results. In particular – as I shall suggest in later chapters – even those neuroethicists who are thoroughly sceptical towards religion, and may imagine their work to be corrosive of theological perspectives, could turn out to be allies in disguise of a Christian ethic – perhaps in unexpected and surprising ways.

Introducing the brain (very, very briefly)

One feature of our brains that enables them to play the central role that they do in our subjectivity, identity and behaviour is their almost unimaginable complexity: it is frequently remarked that the human brain is by far the most complex structure that we know of in the universe. The basic units that make it up, however, are more simple. There are broadly two types of brain cell, neurons (nerve cells) and glial cells. The latter provide physical support and nutrition (among other things) for the neurons, which are responsible for the distinctive functions of the brain. Brain activity consists in electrical signals (in more technical terms, waves of depolarization and repolarization across the cell membrane) that pass along the extensions of neurons known as axons and dendrites. These signals are transmitted from one neuron to another across junctions called synapses, by specific chemical messengers known as neurotransmitters. An electrical signal reaching the end of one axon will cause the release of a neurotransmitter, which diffuses across the synaptic gap and triggers an electrical signal in one of the dendrites of a neighbouring neuron. The complexity arises from the sheer number of these cells and their connections: a human brain contains billions of neurons, each of which may have thousands of synaptic connections with others. The structural organization of the brain is also highly complex, with neurons grouped together into interconnected structures and regions in precise ways to perform the vast number of functions that support the lives we live in the world, from breathing and the beating of our hearts to abstract reasoning and complex communication.[27]

A wide range of approaches have been, and are, used to study brain structure and function in both humans and other animal species.[28] Many of these approaches in various combinations have been used in the neuroscientific studies discussed throughout this book. Anatomical studies of the brain's structure date back at least to the seventeenth century. Much more recently, within the past few decades, techniques like computerized tomography (CT) and especially magnetic resonance imaging (MRI) have been used to generate detailed structural images of living brains. MRI relies on the magnetic properties of the hydrogen nuclei in water, which can be perturbed and detected using a powerful magnetic field. By this means, the distribution of water in various parts of the brain can be detected, and from this, an image of the brain's structure can be reconstructed.

Since the nineteenth century, clinical studies of brain lesions caused by injury or disease have been used to correlate different areas of the brain with psychological and behavioural functions. One early and well-known example was the case of Phineas Gage, referenced briefly in Chapter 3. Brain functions can also be studied by modifying brain activity in various ways. For example,

[27] Space does not permit a detailed account of brain structure and function here, but many good introductory accounts are available. For a brief introduction see, e.g., Antonio Damasio, *Self Comes to Mind: Constructing the Conscious Brain* (London: Vintage, 2012), pp. 299–317, and for a more detailed but very accessible account, Michael S. Gazzaniga, Richard B. Ivry and George R. Mangun, *Cognitive Neuroscience: The Biology of the Mind* (New York: W. W. Norton, 4th edn, 2012), ch. 2. A useful online resource is Anand E. Rughani and Paul L. Penar, 'Brain Anatomy', at http://emedicine.medscape.com/article/1898830-overview (accessed 17 June 2016).

[28] For a history of functional neuroanatomy and a survey of these techniques, see Gazzaniga et al., *Cognitive Neuroscience*, chs. 1 and 3, on which the following overview is based.

some important things have been learned from patients who have had neurosurgery to remove tumours or treat conditions like severe epilepsy. More recently, techniques like transcranial magnetic stimulation (TMS) have been used to induce temporary changes to the activity of specific brain regions and observe their effects on cognition or behaviour. TMS, which will be described more fully in Chapter 6, uses a magnetic field to induce changes to the electrical activity of specific areas of the brain.

Neuronal activity can be studied directly by recording its electrical traces. This is most accurate if the recording electrodes are inserted into the brain or placed on its surface, but the risks and burdens of these procedures severely limit their use with human subjects. The most widely used technique of this sort, therefore, is electroencephalography (EEG), in which the recording electrodes are placed on the subject's scalp and detect electrical signals that pass through the skull. While EEG has very good temporal resolution – being able to detect changes on a timescale of milliseconds – it has much poorer spatial resolution, so that it is often not possible to say with much precision which areas of the brain are generating the signals that are detected.

The techniques that have had a particularly big impact on cognitive neuroscience in recent years – and done much to generate the field of neuroethics – are functional imaging techniques, such as positron emission tomography (PET) and functional magnetic resonance imaging (fMRI). These are based on the assumption that more active regions of the brain will require increased blood flow to supply the oxygen and nutrients needed to support the activity of their neurons. Inferences about regions of the brain involved in mental activities or behaviours can therefore be made by detecting changes in the patterns of blood flow during those activities. FMRI, the most widely used functional neuroimaging technique, relies on the fact that oxygenated blood has different magnetic properties from deoxygenated blood, and so an MRI scanner can be used to detect a blood oxygen level-dependent (BOLD) signal indicating whereabouts in the brain there are raised levels of oxygenated blood.

It is important to keep in mind the limitations of functional neuroimaging techniques. Compared to EEG, fMRI (for example) has a higher spatial resolution but much lower temporal resolution, of the order of seconds rather than milliseconds. Unlike EEG, the BOLD fMRI signal is only an indirect measure of neuronal activity, and the relationship between that activity and the

BOLD signal it generates can be very complex.[29] Also, the raw data generated by fMRI require sophisticated statistical analysis in order to generate information about activity in different regions of the brain, and it is easy for errors and distortions to be introduced in the course of that analysis.[30] Moreover, the design of experiments – particularly in areas as complex as morality or religion – poses both conceptual and practical difficulties. Conceptually, experiments in these areas are inevitably theory-laden, and there is always a danger that the experimental design will be informed by theoretical biases or presuppositions in a way that skews the outcome of the study[31] – a criticism made, for example, of some of the studies of moral judgement discussed in Chapter 3. Practical difficulties include the challenge of studying a large enough sample of participants to give statistically valid results, and the problem known as 'ecological validity' – that is, the impact of the experimental environment itself on the participants' state of mind and thereby on the experimental data. For example, fMRI studies require subjects to be immobilized inside a scanner, which is a large noisy piece of machinery typically housed in a hospital research facility. The experience of prayer or moral deliberation in such an alien environment might be very different from the same activities in their more usual contexts, and those differences are likely to be reflected in the neural activity detected using the scanner.[32]

Introducing this book (also quite briefly)

I asserted earlier that the mutual neglect of neuroethics and Christian ethics is to the detriment of both. This book is an initial attempt to repair that neglect, by exploring a representative range of neuroethical issues from the perspective of a broadly Reformed Protestant tradition. In this way, I hope to make at least an initial contribution to the theological discussion of the

[29] Nikos K. Logothetis, 'What We Can Do and What We Cannot Do with fMRI', *Nature* 453 (2008), pp. 869–78.

[30] Anders Eklund, Thomas E. Nichols and Hans Knutsson, 'Cluster Failure: Why fMRI Inferences for Spatial Extent Have Inflated False-Positive Rates', *Proceedings of the National Academy of Sciences of the USA* 113.28 (2016), pp. 7900–05.

[31] Cf. Adina Roskies, 'Neuroimaging and Inferential Distance', *Neuroethics* 1.1 (2008), pp. 19–30.

[32] Uffe Schjoedt, 'The Religious Brain: A General Introduction to the Experimental Neuroscience of Religion', *Method and Theory in the Study of Religion* 21 (2009), pp. 310–39 (313–15).

actual issues, but also to demonstrate the promise and value of this kind of theological neuroethical work. I hope to persuade Christian ethicists that the questions raised by neuroethics are important, interesting and worthy of their serious attention. Conversely, I hope to show neuroethicists that theological arguments about these questions cannot be ignored, debunked or dismissed as matters of purely private belief. On the contrary, theological reasoning about these questions will have critical and constructive contributions to make, without which neuroethical debate will be the poorer.

As I noted earlier, the contribution of Chapter 2 to achieving these aims is part ground-clearing, part methodological proposal. It is concerned not with a neuroethical topic as such, but with scientific studies of religion. Examining this topic is a way to respond critically to the neuroessentialist and reductive assumptions that (I suggested earlier) help make neuroethics such a secular field. However, it also leads into a critical exploration of the various ways Christian theology might engage with neuroscience. By identifying approaches that seem to me problematic or unsatisfactory, I am able to make my own proposals for how this engagement should be set up in the following chapters.

Across the four chapters concerned with neuroethical topics, there is a general movement from what Roskies calls the 'neuroscience of ethics' to the 'ethics of neuroscience' (though as she also suggests, several of these topics involve interactions between these two aspects). Chapter 3 explores a topic at the heart of the first aspect: the neuroscientific study of morality. Just as neuroscientific studies of religion are sometimes (wrongly) thought to debunk religious faith, so neuroscientific studies of morality are sometimes used to debunk aspects of moral judgement, and to advocate certain kinds of moral theory and approach over against others. In this chapter, I survey some neuroscientific studies of moral judgement and the criticisms levelled at them. I offer my own critique of the normative ethical conclusions their authors try to draw from them. However, I also argue that these authors may prove valuable allies – akin to earlier 'masters of suspicion' such as Feuerbach – in articulating a Christian ethic that in Dietrich Bonhoeffer's words 'claims … to be considered an ethic only as the critique of all ethics'.[33]

[33] Dietrich Bonhoeffer, *Ethics* (Ilse Tödt et al. (eds.); Reinhard Krauss, Charles C. West and Douglas W. Stott (trans.); Dietrich Bonhoeffer Works, 6; Minneapolis, MN: Fortress, 2005), pp. 299–300.

Chapter 4 addresses a range of questions about free will and responsibility raised by neuroscience since the 1980s. I argue that if these questions are relocated in the context of core Christian doctrines, such as providence, sin and salvation, they can be addressed in more satisfactory and interesting ways. In particular, an Augustinian understanding of sin can reframe neuroethical debates about freedom and responsibility in ways that are both fruitful and hopeful.

Chapter 5 is about consciousness and its disorders. The starting point is recent neuroimaging research on vegetative and minimally conscious states, but in order to understand the ethical significance of that research it is necessary to refer to theoretical accounts of consciousness and the self, and bring those into contact with some recent theological discussion about the soul. Returning to practical concerns, I argue that for theological ethics, the presence of conscious awareness matters less and differently than might be supposed in discerning the value and appropriate treatment of patients with disorders of consciousness.

In Chapter 6, I turn to practical questions about the ethics of modifying brain functions technologically, either for therapeutic or enhancement purposes. I offer an approach to the theological analysis of such questions that generates a strong presumption in favour of therapy but a more suspicious attitude to enhancement. However, I argue that what is required in relation to enhancement is not a blanket 'Yes' or 'No', but a theologically informed prudential discernment about each particular project.

This book is an invitation to both Christian ethicists and neuroethicists – as well as others with any interest in either or both fields – to explore what might result from a serious engagement between the two. At the outset, I must beg readers' patience. There is a great deal of science to get to grips with in the areas I discuss, and in each chapter I have found it necessary to give it some detailed attention. I have tried to do the neuroscience the best justice that an outsider to that discipline can, but those with detailed technical knowledge may find areas that I have only partially understood. Those less familiar with neuroscience might find some of this material dense and difficult, though I have tried to present it as clearly as possible. I can only encourage them to persist with it, because it does raise important theological and ethical questions, and if theologians are to address those questions, it is better to do so on the strength of a deep, not superficial, engagement with the science that raises them. In any event, that scientific research has – to my mind – its own intrinsic interest. Both the scientific study of the brain and the theological and ethical questions it raises seem to me to be exciting, challenging and provoking. I hope I manage to convey some of all this in the discussions that follow.

2

Pious Primates with Believing Brains: Evolutionary, Cognitive and Neuroscientific Accounts of Religion

Introduction

In Chapter 1, I speculated that some neuroscientists and neuroethicists might find it easy to dismiss theological perspectives on neuroethics because they assume that scientific studies of religion discredit theology. In this chapter, therefore, I aim to test that assumption by exploring three related approaches to the scientific study of religion and their implications for the validity of the beliefs, practices and experiences they study. These approaches are: first, the cognitive science of religion (CSR), whose aim is to identify cognitive mechanisms that give rise to religious or supernatural beliefs; second, evolutionary theories about the origins of religious beliefs and practices; and third, neuroscientific attempts to identify brain activity and mechanisms associated with religious beliefs, practices and experiences. Together these three offer a range of scientific perspectives that – to echo two recent titles – treat human beings as 'believing primates' with 'religious brains'.[1]

It will come as no surprise to many readers that crude forms of the assumption noted above – that scientific studies of religion discredit religious belief and theology – have little to be said for them. It is widely agreed, even by those very sceptical of religious belief, that scientific explanations of beliefs do not in themselves show those beliefs to be false. This does not mean, however, that scientific studies of religion are irrelevant to theology, much

[1] Jeffrey Schloss and Michael J. Murray (eds.), *The Believing Primate: Scientific, Philosophical, and Theological Reflections on the Origins of Religion* (Oxford: Oxford University Press, 2009); Uffe Schjoedt, 'The Religious Brain: A General Introduction to the Experimental Neuroscience of Religion', *Method and Theory in the Study of Religion* 21 (2009), pp. 310–39.

less to theological engagements with neuroscience. More nuanced forms of scepticism may gain more support from scientific studies than the cruder forms. More specifically, as we shall see, some authors maintain that scientific studies do not *demonstrate* religious beliefs to be false, but may still show them to be epistemically unreliable. While this position itself is certainly not beyond criticism, it does suggest that Christian apologists who rely heavily on scientific evidence to support the credibility of theological claims or determine the shape of their theological accounts might find themselves on shakier ground than they imagine. I shall argue later in the chapter that those wishing to give a theological account of what it is to be human, in dialogue with cognitive science and neuroscience, would do better to let their account be shaped by theological sources and methods, while being open and hospitable to scientific insights critically appropriated according to theological criteria.

It turns out, then, that even quite religiously sceptical readings of the scientific study of religion can perform a valuable critical function for Christian theology, exposing the weaknesses of problematic or unsatisfactory approaches and recalling theology onto firmer ground. In this way even those authors who see themselves as sceptical or hostile towards religion may turn out to be friends in disguise to Christian theology – somewhat as that earlier 'master of suspicion', Ludwig Feuerbach, proved a surprising ally for Karl Barth's theological project.[2]

I should emphasize that the two purposes of this chapter are to clear the ground and gain a methodological sense of direction for the work in the remainder of the book. Therefore, I have not set out to say everything that could be said – or even everything that needs to be said – about cognitive, evolutionary or neuroscientific studies of religion. (Indeed, the treatment of the science in this chapter is somewhat briefer than in some of the subsequent chapters.) A theological engagement with these fields could take a book in itself, but that is not what this book is about. The two headline messages of this chapter are that the scientific study of the brain does not give good reasons to dismiss theological claims or reasoning, and that theological engagements with neuroscience are better framed in a way that gives priority to theological sources and perspectives than by handing over control of the

[2] Karl Barth, *Protestant Theology in the Nineteenth Century* (Brian Cozens and John Bowden (trans.); Grand Rapids, MI: Eerdmans, 2002), ch. 18.

account to scientific perspectives. Readers who need no persuading of these two conclusions could, if they wished, skip this chapter and launch themselves into the discussion of neuroethics proper that begins in Chapter 3. They might, however, still find it helpful to take a brief tour of some of the foundations on which the later chapters of the book are built.

Scientific studies of religious belief and experience

I begin, then, with a brief survey of cognitive, evolutionary and neuroscientific studies of religion.

Cognitive science of religion

In broad terms, cognitive science of religion (CSR) is the attempt to identify cognitive mechanisms that give rise to religious or supernatural beliefs. Its basic premise is that human minds consist of many specific mental 'tools' or 'devices' – cognitive systems for processing different kinds of information.[3] Though CSR accounts are diverse, the field has a widely shared core of ideas that some commentators call the 'standard model'.[4]

The first of these has to do with so-called 'minimally counter-intuitive' (MCI) concepts. These are concepts that violate a few common-sense expectations but not too many: for example, the concept of a human being who can pass through solid walls, but otherwise resembles everyone else, would be MCI. Studies indicate that such concepts are more memorable and easily transmitted than either fully intuitive or highly counter-intuitive concepts.[5] CSR scholars argue that concepts of supernatural agents such as gods are MCI, and therefore tend to persist in our minds and spread in human cultures. Second, we have a 'hypersensitive agency detection device' (HADD): a cognitive system that

[3] These are not necessarily localized in particular regions of the brain: mental modularity does not map straightforwardly onto brain modularity.

[4] Michael J. Murray and Andrew Goldberg, 'Evolutionary Accounts of Religion: Explaining and Explaining Away', in Schloss and Murray (eds.), *The Believing Primate*, pp. 179–99 (183–89). It is worth noting at the outset that one general working assumption is a physicalist or non-dualist understanding of the mind and its relationship to the brain. This will be explored further in Chapter 5.

[5] See, for example, Pascal Boyer, *Religion Explained: The Evolutionary Origins of Religious Thought* (New York: Basic Books, 2001), ch. 2.

attributes events to the intentional activity of agents, even when it has little information on which to base such a judgement.[6] It is argued that this device predisposes us to attribute unexplained events to the activity of supernatural agents.[7] Third, we have a theory of mind, which leads us to attribute minds to other human and human-like beings, and therefore to assume that they have cognitions, beliefs, desires, motivations and so forth rather like ours. This predisposes us to regard supernatural beings as 'minded'.[8] Fourth, concepts of supernatural agents are 'inference rich', enabling our minds to construct narratives about them that enhance their memorability and importance to us.[9] For example, invisible, powerful and all-seeing agents may know what we are up to even when we think no-one is looking,[10] and they could be in a position to affect our lives for good or ill.[11]

CSR is by no means without its critics. One general criticism is that it represents religion in simplistic or caricatured ways. For example, the catch-all category of 'supernatural agents', often used to include ghosts, spirits, zombies, demons, gods and sundry other beings, seems rather distant from the God whom Christians take themselves to encounter in their prayer, worship and service.[12] Sometimes such simplistic and caricatured presentations reflect a sceptical or anti-religious stance.[13] This is not inevitable: Justin Barrett, for example, strongly denies that anti-religious agendas *need* be a part of CSR, though even he acknowledges that it is often used 'rhetorical[ly]' to promote such agendas.[14]

More specifically, the theologian Markus Mühling criticizes CSR for the assumption that the paradigm religious phenomenon is some kind of

6 Justin L. Barrett, 'Cognitive Science of Religion: What Is It and Why Is It?' *Religion Compass* 1.6 (2007), pp. 768–86 (773).

7 Authors vary, though, in the weight they give to the HADD in explaining religious beliefs: compare Boyer, *Religion Explained*, pp. 144–48, with Barrett, 'Cognitive Science of Religion', pp. 772–73.

8 Murray and Goldberg, 'Evolutionary Accounts of Religion', p. 187. Describing theory of mind as *one* cognitive tool may be an over-simplification: see Boyer, *Religion Explained*, pp. 122–23.

9 Murray and Goldberg, 'Evolutionary Accounts of Religion', pp. 187–88.

10 Cf. Dominic Johnson and Jesse Bering, 'Hand of God, Mind of Man: Punishment and Cognition in the Evolution of Co-operation', in Schloss and Murray (eds.), *The Believing Primate*, pp. 26–43.

11 Boyer, *Religion Explained*, pp. 155–60.

12 See further Celia Deane-Drummond, *Christ and Evolution: Wonder and Wisdom* (Minneapolis, MN: Fortress, 2009), pp. 63–68. I would think adherents of various other traditions would find the CSR concept of 'supernatural agents' equally alien, but I write as a Christian theologian and am only really entitled to speak about that tradition.

13 For some examples, see Michael J. Murray, 'Scientific Explanations of Religion and the Justification of Religious Belief', in Schloss and Murray (eds.), *The Believing Primate*, pp. 168–78 (168–69).

14 Barrett, 'Cognitive Science of Religion', p. 779.

exceptional or extraordinary experience. If this assumption is made, he says, a good deal of Christian faith and practice tends to disappear from view.[15] Mühling is surely correct to criticize this assumption, but it is a moot point whether it is a general problem with CSR approaches. Such a view is expressly denied, for example, by Pascal Boyer.[16] The criticism may be nearer the mark in relation to neuroscientific studies, as we shall see later.

To such criticisms, cognitive scientists of religion might reply that CSR only looks for the cognitive mechanisms that generate and support belief; much of the specific *content* of that belief, in particular faith communities and traditions, will be shaped by cultural transmission rather than the underlying cognitive mechanisms.[17] This might partly address the concerns – though Mühling's warning, that a distorted representation of religion may cause much that is important in traditions of faith and practice to disappear from view, remains relevant.

Evolutionary approaches

While CSR is a distinct field from evolutionary theories of religion,[18] the two are closely linked in much of the literature, for fairly obvious reasons.[19] Very broadly speaking, two kinds of approach to explaining religion can be found in the evolutionary literature. One is to propose that religion itself is or was adaptive (conferred a selective advantage) at some time in our species' evolutionary history. The other is to account for it as a non-adaptive by-

15 Markus Mühling, *Resonances: Neurobiology, Evolution and Theology. Evolutionary Niche Construction, the Ecological Brain and Relational-Narrative Theology* (Göttingen: Vandenhoeck and Ruprecht, 2014), pp. 92–97. He makes the same criticism about neurotheology, to be discussed later, though he thinks CSR and neurotheology treat different kinds of exceptional experience as paradigmatic.

16 Boyer, *Religion Explained*, pp. 307–11.

17 Murray and Goldberg, 'Evolutionary Accounts of Religion', p. 189, citing Pascal Boyer, 'Why Do Gods and Spirits Matter at All? Supernatural Gadgets and Social Mind Adaptations', in I. Pyysiäinen and V. Anttonen (eds.), *Current Approaches in the Cognitive Science of Religion* (London: Continuum, 2002), pp. 68–92 (76–77).

18 In a 2007 review Barrett was at pains to distance CSR from evolutionary studies: 'Cognitive Science of Religion', pp. 779–80. However, he has subsequently taken a more positive view of the common ground between them: Justin Barrett, 'Cognitive Science of Religion: Looking Back, Looking Forward', *Journal for the Scientific Study of Religion* 50.2 (2011), pp. 229–39.

19 This is not only true of several contributions to Schloss and Murray (eds.), *The Believing Primate*, but also of some of the standard books in the field such as Scott Atran, *In Gods We Trust: The Evolutionary Landscape of Religion* (Oxford: Oxford University Press, 2002), and Boyer, *Religion Explained*.

product of some other trait or traits that were directly adaptive – a 'spandrel', to use Stephen Jay Gould's term.[20]

Evolutionary explanations that make use of CSR concepts tend to adopt the second of these approaches, treating religion as a spandrel. For example, it is argued, a mental device that enabled our ancestors to recognize other agents with desires and intentions could have conferred an obvious evolutionary advantage on them. Their chances of surviving long enough to reproduce would have been improved by the ability to recognize which objects around them were animals that might regard them as an appetising meal. Moreover, it would have been adaptive for this device to be hypersensitive, because false positives would be less costly than false negatives. Likewise, it is easy to see how the other mental devices in the standard model could have been adaptive in human evolutionary history. The interaction of these various devices might then generate religious concepts and beliefs even though those concepts and beliefs themselves conferred no selective advantage.[21]

Other evolutionary explanations treat religion itself as adaptive, often on the grounds that it promoted group co-operation among our ancestors. One of the more sophisticated accounts of this sort is offered by David Sloan Wilson in his book *Darwin's Cathedral*.[22] Emphasizing that evolution operates at various levels of selection, he gives a group-selectionist account of religions as adaptations for promoting group cohesion. At the heart of his book, a detailed account of John Calvin's Geneva is used as a test-case. Wilson argues that Calvinist belief in all its aspects, as reflected in the first Genevan catechism of 1538, represents a pro-social adaptation. He reads Calvin's catechism as fundamentally concerned with two kinds of relationship: 'people–people' and 'God–people'.[23] Concerning the former, 'Calvin's catechism provides a blueprint for human conduct that makes sense from a group-level functional perspective.'[24] This blueprint includes the Decalogue, the Golden Rule,

20 Stephen J. Gould and Richard C. Lewontin, 'The Spandrels of San Marco and the Panglossian Paradigm: A Critique of the Adaptationist Programme', *Proceedings of the Royal Society B* 205 (1979), pp. 581–98.

21 Boyer, *Religion Explained*, is one good example of this kind of interaction of evolutionary and cognitive science explanations.

22 David Sloan Wilson, *Darwin's Cathedral: Evolution, Religion, and the Nature of Society* (Chicago: University of Chicago Press, 2002).

23 Ibid., p. 94.

24 Ibid., p. 98.

and more contextually specific injunctions about the church and the civic community. Wilson treats Calvin's account of the God–people relationship as 'a fictional belief-system' designed to motivate believers to conform to the requirements of the people–people relationship.[25]

Wilson more or less takes it for granted that Calvin's theology *is* fictional, a point to which I shall return. His way of accounting for theological beliefs also becomes something of a Procrustean bed. He emphasizes aspects of Calvin's theology, like awareness of human depravity and gratitude for God's mercy, which can easily be fitted into his explanation, but more or less ignores other aspects like the Trinity, creation and Christology, which are less convincingly explained as fictions designed to encourage group co-operation.[26] The problem this gives Wilson is that unlike cognitive scientists of religion, he has committed himself to explaining the *whole* package of Calvinist belief and practice, and he has deliberately chosen the 1538 catechism as his dataset. But at best he can only account convincingly for *some* of his chosen data.

A further kind of evolutionary explanation emphasizes cultural rather than genetic evolution. A well-known example of this type is Daniel Dennett's book *Breaking the Spell*.[27] Dennett understands religious beliefs and practices as 'memes', or units of cultural inheritance, which promote behaviours that enable them to persist and spread in the population. The concept of 'memes', however, is itself contentious, and this is just one of the objections that have been levelled at Dennett's account.[28]

Of course, not all these evolutionary explanations need be mutually exclusive. If human evolution operates at multiple levels of selection, it would not be surprising if an area of human life as complex and diverse as religion required a variety of evolutionary (and other) explanations.

[25] Ibid., pp. 98–105.

[26] Wilson's Table 3.2, 'Elements of the God-people relationship specified by Calvin's catechism of 1538' (ibid., p. 101) lists 'Forgiveness of sins', 'Faith', 'Internalization', and 'Preparing for the second coming', but neither the table nor the following discussion even mentions the discussions of the Trinity, creation or Christology found in Article 20 of the catechism, which concerns the Apostle's Creed.

[27] Daniel Dennett, *Breaking the Spell: Religion as a Natural Phenomenon* (New York: Viking, 2006), pp. 179–88.

[28] For critiques of the concept of memes, see Mary Midgley, 'Why Memes?' in Hilary Rose and Steven Rose (eds.), *Alas Poor Darwin: Arguments against Evolutionary Psychology* (London: Jonathan Cape, 2000), pp. 67–84; Jeffrey Schloss, 'Introduction: Evolutionary Theories of Religion. Science Unfettered or Naturalism Run Wild?' in Schloss and Murray (eds.), *The Believing Primate*, pp. 1–25 (24). For other criticisms of Dennett's account, see Armin W. Geertz, 'How *Not* to Do the Cognitive Science of Religion Today', *Method and Theory in the Study of Religion* 20 (2008), pp. 7–21.

The neuroscience of religion

Although the brain's role in religious belief has been of interest at least since the days of William James,[29] the experimental neuroscience of religion is a far younger enterprise: Uffe Schjoedt remarked in a 2009 review that 'the list of actual experiments [was] surprisingly short'.[30] Early hypotheses about the neural bases of religious experience drew on the study of patients with specific brain lesions.[31] For example, it was suggested that temporal lobe epilepsy was associated with hyper-religiosity, and this might indicate a role for the temporal lobes in the generation of religious experience;[32] in journalistic language, they could be the locations of a 'God spot' in the brain. From the late 1980s, Michael Persinger has attempted to test this hypothesis using his 'God helmet', a device claimed to induce the 'sensed presence' of another being by using a weak magnetic field to induce temporal lobe activity.[33] However, it is fair to say that this claim is regarded with considerable scepticism by many in the field.[34]

Various investigators have studied the neural correlates of religious or mystical experiences. Among studies of Christian mysticism, one of the best known was by Mario Beauregard and Vincent Paquette, who carried out fMRI and EEG studies on the mystical experiences of a group of

[29] William James, *The Varieties of Religious Experience: A Study in Human Nature* (New York: Longmans, Green and Co., 1917), online at http://www.gutenberg.org/files/621/621-h/621-h.html (accessed December 3, 2016).

[30] Schjoedt, 'The Religious Brain', p. 333.

[31] For a review, see Patrick McNamara, *The Neuroscience of Religious Experience* (Cambridge: Cambridge University Press, 2009), pp. 80–106.

[32] Michael A. Persinger, 'Religious and Mystical Experiences as Artifacts of Temporal Lobe Function: A General Hypothesis', *Perceptual and Motor Skills* 57 (1983), pp. 1255–62.

[33] For example, L. S. St-Pierre and M. A. Persinger, 'Experimental Facilitation of the Sensed Presence Is Predicted by the Specific Patterns of the Applied Magnetic Fields, Not by Suggestibility: Re-Analyses of 19 Experiments', *International Journal of Neuroscience* 116 (2006), pp. 1079–96.

[34] Schjoedt, 'The Religious Brain', pp. 320–22; for one particular line of critique and Persinger's responses, see: Pehr Granqvist et al., 'Sensed Presence and Mystical Experiences are Predicted by Suggestibility, Not by the Application of Transcranial Weak Complex Magnetic Fields', *Neuroscience Letters* 379.1 (2005), pp. 1–6; M. A. Persinger and S. A. Koren, 'A response to Granqvist et al. "Sensed Presence and Mystical Experiences Are Predicted by Suggestibility, Not by the Application of Transcranial Weak Magnetic Fields"', *Neuroscience Letters* 380.3 (2005), pp. 346–47; Marcus Larsson et al., 'Reply to M.A. Persinger and S. A. Koren's Response to Granqvist et al. "Sensed Presence and Mystical Experiences Are Predicted by Suggestibility, Not by the Application of Transcranial Weak Magnetic Fields"', *Neuroscience Letters* 380.3 (2005), pp. 348–50; St-Pierre and Persinger, 'Experimental Facilitation of the Sensed Presence'. For further critique of Persinger's work, see Mario Beauregard and Denyse O'Leary, *The Spiritual Brain: A Neuroscientist's Case for the Existence of the Soul* (New York: HarperCollins, 2007), ch. 4.

Carmelite nuns.[35] They found that these nuns' experiences were associated with several regions and systems in the brain, arguing against the existence of a single 'God module' or 'God spot'.[36]

Some researchers have proposed particular models or mechanisms for religious experience. For example, in the 1990s, Eugene d'Aquili and Andrew Newberg developed a complex model of the neural mechanisms underlying religious experience.[37] At its heart was a proposed mechanism for the experience of 'absolute unitary being' (AUB), said to be a common feature of mystical experience in diverse traditions. The proposed mechanism involved the blocking of input to the posterior superior parietal lobe, and subsequently Newberg and his co-workers have offered some functional imaging evidence in support of this proposal.[38]

More recently, Patrick McNamara has drawn on a wide range of clinical case reports and brain imaging studies to develop a model of religious experience as self-transformation.[39] He observes that the clinical and brain imaging data consistently implicate large parts of the prefrontal lobes, the anterior temporal cortex and the amygdala in religious experience, and that there is considerable overlap between these areas and those associated with a sense of self.[40] He proposes that the basic problem – cognitive as well as emotional – addressed by religion is 'the divided Self'.[41] In McNamara's model, this problem is solved by

[35] Mario Beauregard and Vincent Paquette, 'Neural Correlates of a Mystical Experience in Carmelite Nuns', *Neuroscience Letters* 405 (2006), pp. 186–90; Mario Beauregard and Vincent Paquette, 'EEG Activity in Carmelite Nuns during a Mystical Experience', *Neuroscience Letters* 444 (2008), pp. 1–4.

[36] Beauregard and O'Leary, *The Spiritual Brain*, p. 274. In this popular work, Beauregard has also argued that neuroscientific studies of religious and mystical experience do not support a materialist view of religion and the brain, but rather support (though they cannot prove) the view that such experiences refer to a divine or supernatural reality beyond the individual subject: ibid., chs. 9, 10.

[37] Eugene d'Aquili and Andrew B. Newberg, *The Mystical Mind: Probing the Biology of Religious Experience* (Minneapolis, MN: Fortress, 1999).

[38] For example, Andrew B. Newberg et al., 'The Measurement of Regional Cerebral Blood Flow during the Complex Cognitive Task of Meditation: A Preliminary SPECT Study', *Psychiatry Research: Neuroimaging* 106 (2001), pp. 113–22; Andrew B. Newberg et al., 'Cerebral Blood Flow during Meditative Prayer: Preliminary Findings and Methodological Issues', *Perceptual and Motor Skills* 97.2 (2003), pp. 625–30; Andrew B. Newberg et al., 'The Measurement of Regional Cerebral Blood Flow during Glossolalia: A Preliminary SPECT Study', *Psychiatry Research: Neuroimaging* 148.1 (2006), pp. 67–71; Andrew B. Newberg et al., 'Cerebral Blood Flow Differences between Long-Term Meditators and Non-Meditators', *Consciousness and Cognition* 19.4 (2010), pp. 899–905; Danny J. J. Wang et al., 'Cerebral Blood Flow Changes Associated with Different Meditation Practices and Perceived Depth of Meditation', *Psychiatry Research: Neuroimaging* 191 (2011), pp. 60–67. For a brief discussion of some of these, see Schjoedt, 'The Religious Brain', pp. 316–18.

[39] McNamara, *The Neuroscience of Religious Experience*.

[40] Ibid., p. xi.

[41] Ibid., p. 23 and ch. 2 passim. There are some interesting echoes here of Christopher Cook's discussion of alcohol addiction and Christian ethics, discussed below in Chapter 4: see Christopher C. H. Cook, *Alcohol, Addiction and Christian Ethics* (Cambridge: Cambridge University Press, 2006).

religious belief and practice through the development of a unified, integrated 'executive Self'; this is achieved by a cognitive process he calls 'decentring', whose goal is to reduce the discrepancy between one's current Self and an 'ideal Self'.[42]

Uffe Schjoedt acknowledges that such models may be useful in generating testable hypotheses. However, he criticizes them for their dependence on conjectures about distinctive brain systems or mechanisms for which there is not yet strong evidence. More generally, he has concerns about the methodology and design of many neuroscientific studies of religion, prior to his own. He is also critical of approaches that focus on exceptional experiences and expert practitioners – echoing from a very different standpoint the criticism by Markus Mühling noted earlier. In his own studies, Schjoedt adopts a piecemeal approach, focuses on everyday experiences and practices – such as different forms of prayer used by lay Christians – and works on the assumption that the neural mechanisms associated with religion will be the same as those involved in other aspects of human life. [43]

Theological implications and a methodological proposal

In the previous section, I surveyed three distinct, though related, scientific approaches to the study of religious belief and experience. Each of these approaches is internally diverse and has its fair share of controversies. There is no shortage of theories and proposals in any of these fields, though fewer seem firmly established. What bearing, though, do these proposals have on either the methods or the content of Christian theology? How can theology best engage with the insights and claims of the evolutionary, cognitive and neurosciences? An answer to these questions will play a central role in shaping the discussion of specific neuroethical topics through the remainder of this book, so it repays some consideration at this point.

The literature on science and theology offers diverse answers. One way to sift through them and draw some constructive conclusions is to use a typology

[42] McNamara, *The Neuroscience of Religious Experience*, pp. 44–58.

[43] For an example of his own approach, see Uffe Schjoedt et al., 'The Power of Charisma – Perceived Charisma Inhibits the Frontal Executive Network of Believers in Intercessory Prayer', *Social, Cognitive and Affective Neurosciences* 6.1 (2011), pp. 119–27.

of approaches that I first developed a few years ago for a different (though related) discussion. (Readers familiar with Hans Frei's *Types of Christian Theology* will recognize my indebtedness to him, though my typology is not simply equivalent to his, having been developed for a different purpose to answer different questions.)[44]

If Christians wish to understand what it is to be human in relation to God, one another and the world, what contribution to that understanding should they expect from theology and what contribution from evolutionary, cognitive and neuroscientific accounts? Broadly speaking, five types of answer can be distinguished: (1) theology makes no contribution to an understanding derived from science alone; (2) both science and theology contribute, but science is the dominant partner in determining the shape and content of the understanding; (3) both science and theology contribute by means of what Frei called *ad hoc* correlation,[45] neither side dominating; (4) both contribute, but theology is the dominant partner, shaping the understanding and critically appropriating insights from science; (5) only theology contributes, science having nothing to offer to a Christian understanding of humanity before God.

Type 5: Only theology contributes, science having nothing to offer to a Christian understanding of humanity before God

To begin with type 5: one form this type can take is Stephen Jay Gould's proposal that science, on the one hand, and religion and ethics, on the other, are 'non-overlapping magisteria' (NOMA):[46] each is valid in its own realm, but they ask distinct and different questions and neither has a bearing on the other. In that case, if Christians wish to understand themselves as moral beings in relation to God, one another and the world, their understanding must be derived entirely from sources 'internal' to the Christian tradition, notably the scriptures and the traditions of faith formed through encounters with those

[44] My typology was first set out in Messer, *Selfish Genes and Christian Ethics*, pp. 49–62, though it is formulated slightly differently here. There will of course also be contributions from other sources, but for simplicity's sake the typology focuses on the relationship between scientific and theological contributions to our human self-understanding. For Frei's typology, see Hans Frei, *Types of Christian Theology* (George Hunsinger and William H. Placher (ed.); New Haven, CT: Yale University Press, 1992), and for an informative discussion, Paul J. DeHart, *The Trial of the Witnesses: The Rise and Decline of Postliberal Theology* (Oxford: Blackwell, 2006), ch. 5.

[45] Frei, *Types of Christian Theology*, p. 38 et passim.

[46] Stephen Jay Gould, *Rocks of Ages: Science and Religion in the Fullness of Life* (London: Vintage, 2002).

scriptures in worship, learning and the life of discipleship. Scientific accounts of religion or morality may be interesting in their own right, but will be simply irrelevant to such a Christian self-understanding.

This position might seem to gain support from the wide consensus that scientific explanations of religion are neutral with respect to religious truth-claims. Most people who work in this area understand well enough that it would be a kind of genetic fallacy to draw conclusions about the truth or falsity of religious beliefs from a causal explanation of how those beliefs come to exist. Even an author as openly hostile to religion as Daniel Dennett can write, 'it could be true that God exists, that God is indeed the intelligent, conscious, loving creator of us all, and yet *still* religion itself... is a perfectly natural phenomenon'.[47] Likewise Paul Bloom draws some familiar parallels with psychological studies of other cognitive functions and beliefs:

> There are psychologists who explore how people come to know that five plus five equals ten, and none of them doubts the truth of five plus five equals ten... Psychologists who study why people believe there is intelligent life on Mars would be very confused if they thought their findings would bear on the debate over the actual existence of extraterrestrial life.[48]

Such a division of labour might initially seem an attractive way to keep the peace between science and theology, but in the fields we are considering, the border between Gould's two 'magisteria' turns out to be difficult, maybe impossible, to mark out and police.

One aspect of the difficulty is that, as Jeffrey Schloss observes, neutrality is not always what it seems.[49] In some cases, it may indeed reflect a commitment to strict objectivity in the scientific study of religion, and a concern that *either* pro- *or* anti-religious agendas may compromise that objectivity. Thus, Armin Geertz, for example, is hostile to neurotheology because he discerns in it a religious agenda that risks compromising the scientific quality of its studies. He exhorts neuroscientists who study religion: 'Please keep your religious

[47] Dennett, *Breaking the Spell*, p. 25 emphasis original.

[48] Paul Bloom, 'Religious Belief as an Evolutionary Accident', in Schloss and Murray (eds.), *The Believing Primate*, pp. 118–27 (125).

[49] Schloss, 'Evolutionary Theories of Religion', pp. 10–12. For further exploration of difficulties with neutrality, see Dominic D. P. Johnson, Hillary L. Lenfesty and Jeffrey P. Schloss, 'The Elephant in the Room: Do Evolutionary Accounts of Religion Entail the Falsity of Religious Belief?' *Philosophy, Theology and the Sciences* 1.2 (2014), pp. 200–31.

agendas out of your work… Otherwise we can only conclude that your results indicate nothing other than teleological confirmation of religious claims.'[50] But he is equally critical of the anti-religious polemic in Dennett's *Breaking the Spell*.[51]

In others, though, neutrality goes along with a general scepticism towards religious truth-claims. Paul Bloom, for example, maintains that while scientific explanations cannot *refute* theistic beliefs, they can nonetheless expose those beliefs as epistemically unreliable.[52] In short, many evolutionary and cognitive scientists of religion, though they profess some kind of neutrality, are better described by type 1 than type 5.

Moreover, for a faith that regards the physical world and its living inhabitants as God's good creation, there are theological reasons for being uneasy with the NOMA solution. Even if we say, as I believe we should, that the scientific study of nature will not teach us that the world is – *or* that it is not – God's creation, it would be odd if science had nothing of importance to teach us about the world that we affirm, on theological grounds, to *be* God's creation. Indeed, according to Colin Gunton, the Christian doctrine of creation provides a strong theological motivation for natural science: because the world is God's good creation, it is worth studying; because it is contingent rather than necessary, we shall only find out what it is like by empirical investigation, not by *a priori* reasoning alone.[53]

Type 1: Theology makes no contribution to an understanding derived from science alone

At the other end of the scale lies the view I have characterized as type 1. This is hardly a promising option for those wanting to articulate a *Christian* understanding of what it is to be human, since it is in effect a rejection of that project: a call for theological understandings to be displaced by those based on science alone. However, it requires some evaluation, because such a

[50] Armin W. Geertz, 'When Cognitive Scientists Become Religious, Science Is in Trouble: On Neurotheology from a Philosophy of Science Perspective', *Religion* 39.4 (2009), pp. 319–24 (324).

[51] Geertz, 'How *Not* to Do the Cognitive Science of Religion Today'.

[52] Bloom, 'Religious Belief as an Evolutionary Accident', p. 125.

[53] See Colin Gunton, *The Triune Creator: A Historical and Systematic Study* (Grand Rapids, MI: Eerdmans, 1998), pp. 102–16.

view is quite frequently promoted (explicitly or implicitly) in the literature on scientific approaches to the study of religion, in which some form or other of naturalism is commonplace.

Now there are some forms of naturalism with which Christian theology need have no quarrel. It is widely (though not quite universally) agreed that natural scientific research is quite properly *methodologically* naturalist,[54] aiming to account for natural phenomena by natural cause and effect, without invoking supernatural causes to fill explanatory gaps. There might also be *theological* reasons for thinking that science should be done in a methodologically naturalist way: in a particular and limited sense, *etsi deus non daretur*. To introduce divine intervention as a scientific explanation of natural phenomena (as Newton did, and Laplace did not, to explain the motions of the planets) risks eliding the distinction between divine and creaturely causation to the extent that we are no longer speaking of the God whom Christians worship, but a demiurge – a demiurge of the gaps, moreover, who becomes less and less necessary as scientific explanations become more complete.[55] Presumably, if explaining the existence of religious belief and practice and the forms they take is a valid area of natural-scientific enquiry, it should be no exception to the rule that natural science is best done in methodologically naturalistic ways. But it is a further step from this methodological naturalism to the *ontologically* naturalist view that there is no God and (as Peter van Inwagen puts it) '[n]aturalistic explanations of supernaturalistic belief… are "all the explanation there is"'.[56] In the literature, the latter claim may be related to methodologically naturalistic studies of religion in one of two ways.

For some authors, it is an implicit or explicit prior assumption. It seems to function in this way for David Sloan Wilson, whose detailed study of Calvin's

[54] For a minority voice arguing that genuine science need not be methodologically naturalist, see Alvin Plantinga, 'Games Scientists Play', in Schloss and Murray, eds., *The Believing Primate*, pp. 139–67.

[55] As my use of the phrase *etsi deus non daretur* hints, there is some connection here with Dietrich Bonhoeffer's reflections in the prison letters on the abandonment of 'God as a working hypothesis', and his well-known dark phrase, 'The God who lets us live in the world without the working hypothesis of God is the God before whom we stand continually. Before God and with God we live without God'. Dietrich Bonhoeffer, *Letters and Papers from Prison* (Eberhard Bethge (ed.); Reginald Fuller et al. (trans.); London: SCM, 1971), p. 360. For an illuminating reflection on the context and significance of the *etsi deus* in Bonhoeffer's prison writings, see Kevin A. Lenehan, '*Etsi Deus Non Daretur*: Bonhoeffer's Useful Misuse of Grotius' Maxim and its Implications for Evangelisation in the World Come of Age', *The Bonhoeffer Legacy: Australasian Journal of Bonhoeffer Studies* 1.1 (2013), pp. 34–60.

[56] Peter van Inwagen, 'Explaining Belief in the Supernatural: Some Thoughts on Paul Bloom's "Religious Belief as an Evolutionary Accident"', in Schloss and Murray (eds.), *The Believing Primate*, pp. 128–38 (134).

Geneva in *Darwin's Cathedral* is offered in support of his hypothesis that religion is adaptive because it promotes group co-operation and cohesion. In other words, Wilson asks a methodologically naturalist question about Calvinism, and not unexpectedly, he partially succeeds in getting a naturalist answer. (As I noted earlier, though, there are parts of his chosen dataset, Calvin's 1538 catechism, which resist assimilation to his naturalist explanation, and which he more or less ignores.) The success of his methodologically naturalist account, however, does not warrant his ontologically naturalist assumption that Calvinist beliefs about God are fictitious. Nor does it warrant the 'demolition with respect to theism' that he attempts against Christian Smith's book *Moral, Believing Animals*.[57] Against Smith's claim that a theistic explanation of religious origins, while not empirically verifiable, is consistent with a plausible anthropological account,[58] Wilson insists that religious beliefs *are* empirically verifiable but have been falsified by Darwinian evolutionary theory. But his argument is unconvincing: he cites an odd-looking list of examples as evidence,[59] and displays a surprising lack of sophistication, for one who has previously demonstrated some thoroughness and care in his analysis of Christian faith and practice, in understanding what a Christian tradition might mean by belief in God. In short, Wilson fails to make a persuasive case for shifting from a purely methodological to an ontological naturalism. This means that Christians might find illuminating insights in his account of Christianity as adaptive (though as I have suggested, his account is not without explanatory weaknesses); but if they do, they need not agree that his scientific explanation excludes other, theological, ones.

57 Christian Smith, *Moral, Believing Animals: Human Personhood and Culture* (Oxford: Oxford University Press, 2003); David Sloan Wilson, 'Evolutionary Social Constructivism: Narrowing (but Not Yet Bridging) the Gap', in Schloss and Murray, eds., *The Believing Primate*, pp. 319–38 (quotation at 338).

58 Smith, *Moral, Believing Animals*, pp. 95–123.

59 'For example, it is perfectly possible empirically for the earth to be young, for each species to be separately created and unchanging through time, for organisms to be elegantly designed in every respect, and for life's afflictions to be part of a benign grand plan when understood in greater detail. These claims *could* be true, but they are disconfirmed by the empirical evidence.' Wilson, 'Evolutionary Social Constructivism', p. 335, emphasis original. The first is not explicitly a religious claim, though it was generally assumed by Christians before the advent of modern geology; the second is one possible reading of Genesis 1, but by no means the only one; the third comes from a very particular time and place (eighteenth and nineteenth century English natural theology) and would not have been made in this way, for example, by a premodern theologian like Thomas Aquinas; it is not clear that the final one really is an empirically testable claim, but it has in any case been thoroughly problematized *within* Jewish and Christian traditions since the writing of the book of Job.

For others, ontological naturalism seems to be a conclusion that they think their studies suggest or support. Few of course are naïve enough to suggest that scientific explanations of religious beliefs *refute* those beliefs: they understand perfectly well that this would be a genetic fallacy, as Bloom makes clear in the passage quoted earlier.[60] It is more often thought, however, that scientific explanations show religious beliefs to be epistemically dubious in some way – to use van Inwagen's phrase, that such explanations '*resist* being incorporated into a larger, more comprehensive supernaturalistic explanation'.[61] Michael Murray considers four reasons for thinking this: the mechanism generating religious beliefs is prone to false positives, the various beliefs generated by natural mechanisms are mutually exclusive, religious beliefs do not have a proper causal connection to the object of those beliefs, and supernatural beliefs have not been subjected to the 'winnowing power of natural selection' in the same way as (for example) beliefs about the correspondence between our visual perceptions and the physical world.[62] Though Murray rejects these arguments as unsound, they are widespread in the literature. For example, having acknowledged that scientific explanations cannot refute theistic beliefs, Bloom goes on to argue that they could nonetheless 'challenge the rationality of those who hold such beliefs' by showing that the route to those beliefs is epistemically unreliable.[63]

What, though, does this epistemic challenge to belief really amount to? At most, as Gregory Peterson has argued, it is a 'hermeneutics of suspicion'.[64] In other words, if there are neuropsychological factors that motivate us to hold certain beliefs, this might lead us to suspect that the reasons we give for those beliefs could really be rationalizations. It could be that our reasons for believing seem persuasive to us, not because they really *are* good reasons, but because we have psychological mechanisms that motivate us to *find* them persuasive. This 'hermeneutics of suspicion' hardly justifies ontological naturalism, but it might turn out to have some theological uses of its own, as we shall see.

With the two extreme positions in the typology ruled out, the three intermediate types remain to be considered.

[60] Bloom, 'Religious Belief as an Evolutionary Accident', p. 125.

[61] Van Inwagen, 'Explaining Belief in the Supernatural', p. 135, emphasis original.

[62] Murray, 'Scientific Explanations of Religion', passim (quotation at p. 177).

[63] Bloom, 'Religious Belief as an Evolutionary Accident', p. 125.

[64] Gregory R. Peterson, 'Are Evolutionary/Cognitive Theories of Religion Relevant for Philosophy of Religion?' *Zygon* 45.3 (2010), pp. 545–57.

Type 2: Both science and theology contribute, but science is the dominant partner in determining the shape and content of the understanding

Any apologetic strategy that relies heavily on scientific evidence to make the case for the existence of God or spiritual reality can be considered an instance of type 2, because to a large extent the theological claims will stand or fall by the scientific data and their interpretation, and scientific evidence is likely to have a big influence on *what* theological claims are made.[65] For example, Mario Beauregard maintains that a range of scientific evidence, including his own studies of Christian mystical experience, cannot prove but does support 'the view that individuals who have [religious, spiritual and mystical experiences] do in fact contact an objectively real "force" that exists outside themselves'.[66] The neurotheology developed by d'Aquili and Newberg in the 1990s also has an apologetic element in that it offers a neuroscientific explanation of 'why God won't go away'.[67] While they concede that it is 'in principle unknowable' whether there is any objective referent of religious experience outside the brain, they hold that this is also true of the 'baseline reality' of everyday experience, and argue that absolute unitary being has a claim to be considered 'ultimate reality': it would be 'foolish reductionism' to insist that 'because hyperlucid unitary consciousness can be understood in terms of neuropsychological processes, it is therefore derivative from baseline reality'.[68]

The neurotheology of *The Mystical Mind* goes further than apologetics, however, offering an ambitious programme for the reform and unity of the world's diverse religious and theological traditions. D'Aquili and Newberg propose that their neuroscientific account of religious belief can account for how and why 'foundational myths' are generated in any belief system, how and why these are 'elaborated into complex logical systems that we call specific

[65] Among the leaders of the science and theology field in the second half of the twentieth century, for example, this was true of Arthur Peacocke, who was convinced that to recover any intellectual credibility, theology must reform itself to conform to the natural sciences both in its methods and in its substantive content. For a concise statement of his position, see Arthur Peacocke, 'Science and the Future of Theology: Critical Issues', *Zygon* 35.1 (2000), pp. 119–40. Forms of apologetics in which science plays a less determinative role in making the case for Christian belief and determining its content might fit more easily into type 3.

[66] Beauregard and O'Leary, *The Spiritual Brain*, pp. 289–90.

[67] Eugene G. d'Aquili and Andrew B. Newberg, 'The Neuropsychological Basis of Religions, or Why God Won't Go Away', *Zygon* 33.2 (1998), pp. 187–201.

[68] Ibid., p. 200.

theologies', and how and why the myths and theologies generate 'the motor behaviour that we call ceremonial ritual'.[69] This means that the neuroscientific study of religion can supply a 'metatheology', by which they mean 'the general principles describing and, implicitly, the rules for constructing any concrete theological system' in any of the world's religions or belief systems.[70] (Suspicious readers might discern some sleight of hand here: the sentence just quoted shifts from descriptions and causal explanatory models of the sort supplied by neuroscience to normative claims about how theological systems *should* be constructed, but it is not made clear how the theological 'ought' might be derived from the scientific 'is'.) Moreover, they maintain that the neuroscience of religion can supply not just a metatheology, but a 'megatheology'. The latter

> should contain content of such a universal nature that it could be adopted by most, if not all, of the world's great religions as a basic element without any serious violation of their essential doctrines. Alternatively, a megatheology should have such universal content that it could be used as the basis for the development of a new specific theology, one, it is to be hoped, more universal in nature than those arising from the cultural exigencies of humanity's remote past.[71]

This looks like the rather bold claim that neuroscience will provide something akin to what the Reverend Mr Casaubon, in George Eliot's *Middlemarch*, called the 'Key to all Mythologies',[72] with the potential to unite the diverse religious traditions and theologies found in the world today. D'Aquili and Newberg suggest this can be achieved through the study of absolute unitary being, since 'according to the only criterion by which we can judge reality, AUB is more "really real" than any other known phase of consciousness. At this point, one cannot refrain from asking, "Are we looking at God?"' They push their speculations further: AUB may be experienced in two forms, one 'suffused with positive affect', which is generally experienced as personal and referred to as 'God' by those who experience it, and the other 'suffused with neutral affect', experienced as non-personal and referred to as 'void consciousness'. D'Aquili and Newberg propose that we should see the second of these as 'the anterior

[69] D'Aquili and Newberg, *The Mystical Mind*, p. 195.
[70] Ibid.
[71] Ibid., p. 198.
[72] George Eliot, *Middlemarch* (W. J. Harvey (ed.); London: Penguin, 1994), p. 87.

nature of God', representing 'a total and infinite conscious potentiality', while the first form of AUB should be 'the posterior nature of God', or 'total actuality'; these two concepts 'have the virtue of being able to unite God seekers and void consciousness seekers in one superordinate conceptual framework'.[73]

Well, perhaps. Those trained in the methods and content of Christian theology may find such proposals vague, speculative and shot through with what Karl Barth called 'abstraction'; but perhaps that is only to be expected in such a new and untried enterprise. Another difficulty for type 2 projects of the apologetic or the more ambitious kind, however, is suggested by Peterson's hermeneutic of suspicion, because scientific studies of religion not only offer naturalistic explanations of religious phenomena; they also suggest naturalistic reasons why we will be predisposed to find non-naturalistic explanations of those phenomena more plausible. In other words, a theology that relies heavily on science to support its claims and to guide its methods and content will risk being exposed as epistemically unreliable in the way described by Paul Bloom.[74] It will then be in danger of collapsing into type 1, in which methodologically naturalistic scientific accounts of religion generate the ontologically naturalist claim that scientific accounts are 'all the explanation there is'. On closer inspection, then, type 2 does not seem such a promising strategy for theological engagement with evolutionary, cognitive or neuroscientific research.

Moreover, by enlisting science to do theological work, this type of engagement may also hinder science from doing its own proper work. If science is indeed best done in methodologically naturalistic ways, then using it to support non-naturalistic claims or investigate non-naturalistic questions risks distorting either the scientific investigation or the answers it supplies. In that sense, Geertz may be right to insist that religious agendas should be kept out of scientific research, with the proviso that this also applies to atheistic agendas: these too are, in a negative sense, religious agendas that science *qua* science is not competent to support.[75] Ironically, a type of theological

[73] All quotations in this paragraph are from D'Aquili and Newberg, *The Mystical Mind*, p. 201. In his more recent *Principles of Neurotheology*, Newberg appears to have rowed back somewhat from these bold claims, and indeed at one point maintains that neuroscience and theology 'should have similar, and reciprocal, emphasis in the overall [neurotheological] dialogue' (p. 54), a position that might seem closer to my type 3. However, he still thinks neurotheology should be a metatheology and a megatheology (ibid., pp. 64–66), and such aspirations locate his neurotheology firmly in type 2.

[74] Bloom, 'Religious Belief as an Evolutionary Accident', p. 125.

[75] Geertz, 'When Cognitive Scientists Become Religious'.

engagement like type 4, which sets narrower limits on the part natural science should play in shaping a Christian understanding, may thereby show *more* respect for its integrity as an intellectual practice.

Type 4: Both science and theology contribute, but theology is the dominant partner, shaping the understanding and critically appropriating insights from science

This brings us to type 4, and it is worth recalling again Peterson's scientific hermeneutic of suspicion. Like others, he makes the connection with earlier 'masters of suspicion' (as Ricoeur called them), remarking that the cognitive science of religion is often 'seen as an inheritor of the Feuerbachian project to reduce religion to something else, whether it be psychology, class struggle, or will-to-power'.[76] Christian theology, of course, has been here before. Although the Feuerbachian critique has often been thought to threaten Christian faith, it is well known that for Karl Barth (for one), Feuerbach turned out to be an ally.[77] By exposing the human tendency to project our own experience onto a transcendent screen and call it 'God', a Feuerbachian type of critique forces us to acknowledge that 'one can *not* speak of God simply by speaking of man in a loud voice'.[78] Barth made this famous remark about Friedrich Schleiermacher in an early essay, and there is a growing consensus that it is not only unfair to Schleiermacher, but unrepresentative of Barth's mature evaluation of him.[79] However, it seems all too apt an assessment of the kind of neurotheology which claims to learn from the scientific analysis of different mystical experiences that God has 'anterior' and 'posterior' natures.[80] Against such transcendental projections of human experience, Christian theology is recalled to the only ground on which it is entitled to stand, understanding itself as a faithful and rigorous response to God's self-disclosure in Christ.

If that is the central task of Christian theology, how should it interact with the natural sciences in articulating an understanding of what it is to be human before

76 Peterson, 'Evolutionary/Cognitive Theories of Religion', p. 547.

77 Barth, *Protestant Theology in the Nineteenth Century*, ch. 18.

78 Karl Barth, *The Word of God and the Word of Man* (Douglas Horton Smith (trans.); New York: Harper, 1957), p. 196.

79 See, e.g., Bruce L. McCormack, *Orthodox and Modern: Studies in the Theology of Karl Barth* (Grand Rapids, MI: Baker Academic, 2008), ch. 3.

80 D'Aquili and Newberg, *The Mystical Mind*, p. 201.

God and in the company of others? Barth himself might seem an unpromising example, since he sometimes appears to deny that the natural sciences have any contribution to make to Christian doctrine. In the preface to the first part of his doctrine of creation, for example, he anticipates Gould's NOMA:

> There is free scope for natural science beyond what theology describes as the work of the Creator. And theology can and must move freely where science which really is science… has its appointed limit. I am of the opinion, however, that future workers in the field of the Christian doctrine of creation will find many problems worth pondering in defining the point and manner of this twofold boundary.[81]

This appears to restrict the interaction of theology with natural science to the negotiation and policing of borders. However, elsewhere in his work Barth shows himself quite ready to engage creatively with non-theological disciplines, and to appropriate their insights critically for his theological work,[82] so there seems no reason in principle why a theological investigation done somewhat after the manner of Barth cannot engage critically and constructively with evolutionary, cognitive and neuroscientific perspectives on human nature.[83]

Type 3: Both science and theology contribute by means of '*ad hoc* correlations', neither side dominating

Type 3 remains to be considered. While this type of theological engagement resists any dominance of science in shaping a Christian understanding of humanity, and thus allows space for theology to operate according to its proper sources and methods, I have in the past questioned how stable it is. I have speculated that this type of engagement is always liable to drift towards type 2 or 4, so that either science or theology comes to dominate the encounter, and the real methodological choice in science and theology is between those

[81] Karl Barth, *Church Dogmatics* (G. W. Bromiley and T. F. Torrance (ed.), 13 vols., Edinburgh: T&T Clark, 1956–75), vol. 3.1, p. x.

[82] Nigel Biggar, *The Hastening that Waits: Karl Barth's Ethics* (Oxford: Clarendon Press, 1993), ch. 5.

[83] For further reflections on Barth and scientific understandings of human nature, see Philip Chapman, 'Barth and Darwin: What Is Humanity?', *Theology and Science* 12.4 (2014), pp. 362–77. On Barth and science more generally see also McCormack, *Orthodox and Modern*, ch. 12.

two types.[84] Moreover, in this field, a drift to type 2 is perhaps the more frequent temptation. This is not to say, however, that it always happens. For example, the recent book by Markus Mühling cited earlier offers a highly sophisticated and genuinely mutually critical encounter between theology, phenomenology, neuroscience and evolutionary theory, which does seem to avoid the temptation.[85]

What conclusions does all this suggest about how theology should engage with science in the neuroethical arena? Although the typology outlined here is not simply equivalent to that of Hans Frei, some recent reflections by Paul DeHart on Frei's typology may by analogy prove instructive for mine.[86] On DeHart's reading, Frei's third and fourth types (*ad hoc* correlation and theological subordination of external discourses) represent the range of legitimate possibilities for theology, and they form what he calls a 'mutually stabilizing pair',[87] enabling the third to remain clearly distinct from the second type and keeping the fourth from drifting towards 'the theological dead end represented by Type Five'.[88]

Something comparable could be said about my types 3 and 4.[89] I have suggested that when science plays a dominant role in shaping a Christian understanding of human being (type 2), this may have a distorting effect on the science as well as the Christian understanding. I have also suggested that type 3 is often at risk of drifting towards type 2. On the other hand, a hazard for type 4 is to drift towards type 5, separating theological and scientific discourses and denying any relevance of one to the other. Barth's remarks about the 'twofold boundary', which I quoted a moment ago, are a case in point. Perhaps, therefore, my types 3 and 4 also make a 'mutually stabilizing pair', in DeHart's phrase, each drawing the other away from its characteristic pitfall. If so, Christian theological dialogue with the natural sciences needs both types; but historically type 3 has arguably been a more mainstream approach to science and theology than type 4. One reason for attempting a type 4 engagement in this book is to redress that balance: to test the capacity of a Christian theology

[84] Messer, *Selfish Genes*, pp. 59–60.
[85] Mühling, *Resonances*.
[86] DeHart, *The Trial of the Witnesses*, pp. 205–17.
[87] Ibid., p. 217.
[88] Ibid., pp. 216–17.
[89] This represents a shift of emphasis from my previous use of the typology in *Selfish Genes*, pp. 49–62.

firmly rooted in its own distinctive sources and methods to appropriate insights critically from evolutionary, cognitive and neuroscientific research, to incorporate them in a theologically-formed vision, and so to articulate a distinctive and compelling theological neuroethics.

In making this proposal, I have concentrated on the engagement between theology and the natural sciences, since it will be crucial to get this relationship right in developing a theological neuroethics, and – as I have suggested in this chapter – it seems all too easy to get it wrong in various ways. But of course there are other important interdisciplinary engagements to be considered in developing a theological neuroethics, most notably with philosophical ethics. For the present, suffice it to say that the engagement with philosophical approaches to neuroethics in the remainder of this book will be conducted in an analogous manner to what I have advocated for a theological engagement with the natural sciences: by engaging as carefully and attentively as possible with philosophical debates, and also being ready to reframe those debates along theological lines, and appropriate philosophical insights critically into a theologically shaped account.

Conclusion

The purposes of this chapter were to clear the ground for a theological engagement with neuroethical topics in the remainder of the book, and to suggest how that engagement might best be carried out. By considering some questions raised for Christian theology by evolutionary, cognitive and neuroscientific studies of religion, I have argued that such studies do not give good grounds for an ontologically naturalist view of the brain and mind. Even if ontologically naturalist views are widespread among neuroethicists, they are not well-founded and do not give good grounds to ignore or dismiss the contribution theology can make to neuroethics.

Theological reflection on scientific studies of religion has also suggested some conclusions about how theology should engage with the natural sciences, and therefore how that theological contribution to neuroethics should be made. I have argued that approaches which give the natural sciences too great a role in determining the methods and content of theological reasoning do a disservice not only to theology but also to the natural sciences. The proper limits to the range of ways to conduct a science-theology dialogue, I have suggested, are set by types 3 and 4 in my typology: *ad hoc* correlations in which neither side dominates, and encounters in which the agenda is set by theology and scientific insights are critically appropriated according to theological criteria. It may turn out (as I suggested with reference to Paul DeHart) that theology needs both of these types of approach in its engagement with the natural sciences, because each can steer the other away from its own characteristic dangers and distortions. In any event, since type 4 approaches have historically been under-represented in theological engagement with the natural sciences, that is, one good reason for adopting this kind of approach.

That is my aim in this book. Each of the following chapters addresses a specific neuroethical topic, beginning with critical surveys of the relevant science and philosophical-ethical responses to it. These surveys may be quite extensive, because both the science and the philosophical discussions can be voluminous, and it is important to engage carefully with them. But when we move to theological analysis in the later parts of each chapter, the emphasis is on relocating the perspectives and questions from the scientific and philosophical discussions in a frame of reference shaped by the core convictions of Christian faith. In that frame not only the answers but also the questions may look rather different. Theological neuroethicists may find that they need to frame the issues differently from their philosophical and scientific colleagues, as well as offering distinctive responses. This might sometimes complicate neuroethical debates; but I believe it will complicate them in fruitful, interesting and illuminating ways.

3

'Like God, knowing good and evil': The Neuroscience of Morality and the Theological Suspicion of Ethics

Introduction

No topic better illustrates both the promise and the pitfalls of neuroethics than the neuroscience of morality. What is going on in our brains when we make moral decisions? What systems and structures of our brains are involved in moral judgement, and what parts do they play? Does this tell us anything useful about the potential and the limitations of our moral cognition? Should we try to draw normative conclusions from this, or would that simply represent some kind of conceptual confusion?

One of the most prominent – and controversial – figures in this field is Joshua Greene. In the early 2000s, he began a series of functional imaging studies investigating subjects' brain activity while they considered the moral thought-experiments known as 'trolley problems'. These involve scenarios where acting to save several lives will result in one person's death, but they vary in the connections between the saving of several lives and the one death. On the basis of his imaging studies and earlier psychological work by Jonathan Haidt, Greene argues that non-utilitarian moral judgements about such scenarios reflect automatic, intuitive decision-making processes in the brain that are the product of our evolutionary history. Deontological moral *theories*, in his view, reflect the use of reason to rationalize these intuitive judgements after the fact. Much of the time, Greene thinks, our moral intuitions serve us well, but in the complexities of modern life they can lead us into intractable conflicts that can only be resolved by utilitarian reasoning.

More or less every aspect of Greene's work, from the conceptualization and design of his experiments to his philosophical reasoning, has come in for criticism from others. In this chapter, I shall survey his own arguments and many of these criticisms. However, the main purpose of the chapter is to explore how a Christian theological ethic might engage with these debates, and to this I shall propose an answer that might initially seem surprising. I shall argue that if Greene's work (for all its problems) does tend to deflate our estimate of the power of human moral reason, he may prove a friend in disguise to Christian ethics. To be more specific, he might prove an unexpected ally to the theological suspicion of the human project of ethics expressed by theologians such as Bonhoeffer and Barth – which for them was a necessary first step in the theological reconstruction of ethics. In this reconstructive work, however, theologians will have every reason to reject Greene's more tendentious arguments. They can welcome his critical insights (to the extent that these stand the test of time) without being committed to his normative conclusions.

Perspectives from evolution and moral psychology

Greene's neuroscientific work builds on earlier work in moral psychology, and the conclusions he draws from his studies rely (among other things) on claims about the evolutionary origins of morality. Therefore, before we begin to explore Greene's studies and the ethical argument he builds on them, this section offers a very brief account of that background. I begin with evolutionary arguments about morality, and then outline the theory of moral psychology that has most directly influenced Greene. Along the way I shall note some distinctive features of these evolutionary and psychological discussions, which also have a bearing on Greene's work and how it should be assessed.

Ever since Darwin published the *Origin of Species*, evolutionary biologists have sought to explain the origins of human morality. Often, the heart of this problem has been taken to be the existence of altruism: that is, behaviour by one individual that improves others' chances of survival and reproduction at the expense of its own. Altruistic behaviour so defined is widespread in the animal kingdom, from eusocial insects such as honey bees to various bird and mammal species. As Darwin acknowledged, it is a serious puzzle for the theory of natural selection, because behavioural traits that lead individuals to diminish their own chances of survival and reproduction for the sake of other group members could be expected to disappear from the population over time.[1] Two elegant, if partial, theoretical solutions were proposed in the 1960s and after. The first was that some altruistic behaviour benefits close kin, so that even if altruistic individuals die or fail to reproduce, their self-sacrifice may help the genes they share with their relatives to persist and spread in the population. The second was that other instances of altruistic behaviour are reciprocal (if you scratch my back, I'll scratch yours), and improve both parties' reproductive fitness.[2]

Humans, however, display 'genuine', non-reciprocal, non-kin altruism – put simply, behaviour such as kindness to strangers – which is more difficult to explain in terms of evolutionary selection.[3] Some theorists have argued that genuine altruism confers no selective advantage, but is a by-product of other behavioural traits that were adaptive in our evolutionary history. Others regard it as a form of manipulation: altruistic individuals are in effect duped into serving the interests of others, somewhat as host birds are duped into looking after cuckoo eggs and fledglings. Still others argue that genuine altruism and other 'groupish' behaviours are the products of cultural, not biological, evolution. The problem becomes easier if one allows that natural selection can operate on groups as well as individual organisms or their genes.[4] In that case, genuine altruism (like other co-operative behaviours that may benefit the group at the expense of the individual) can persist and spread if it gives the group a selective advantage over other groups. For group selection to work, means of enforcing group co-operation are needed; as we saw in the

1 Charles Darwin, *The Origin of Species* (1859; repr. New York: Gramercy, 1979), pp. 257–62.

2 William D. Hamilton, 'The Genetical Evolution of Social Behaviour', parts 1 and 2, *Journal of Theoretical Biology* 7 (1964), pp. 1–52; Robert L. Trivers, 'The Evolution of Reciprocal Altruism', *Quarterly Review of Biology* 46 (1971), pp. 35–56.

3 On the following proposals, see further Neil Messer, *Selfish Genes and Christian Ethics: Theological and Ethical Reflections on Evolutionary Biology* (London: SCM Press, 2007), pp. 67–70, and references therein.

4 This was rejected by almost all evolutionary theorists from the 1960s until the 1990s, but since the 1990s some, such as David Sloan Wilson, have attempted to rehabilitate group-selection theory: see Elliot Sober and David Sloan Wilson, *Unto Others: The Evolution and Psychology of Unselfish Behavior* (Cambridge, MA: Harvard University Press, 1998); Christopher Boehm, 'Explaining the Prosocial Side of Moral Communities', in Philip Clayton and Jeffrey Schloss (eds.) *Evolution and Ethics: Human Morality in Biological and Religious Perspective* (Grand Rapids, MI: Eerdmans, 2004), pp. 78–100.

previous chapter, David Sloan Wilson and others have argued that religion can be accounted for as just such a pro-social behaviour.[5]

Two features of this discussion are worth noting. First, as observed earlier, explaining altruism is often taken to be the central task, because the core of morality is assumed to be what Auguste Comte called 'the subordination of Egoism to Altruism'.[6] But it is not always appreciated how historically and culturally specific this assumption is. Both the term 'altruism' and this way of framing the central problem of morality are generally credited to Comte. Alasdair MacIntyre has argued that in the premodern West, the view of human beings implicit in this framing of the issue, as individuals whose goods are always potentially in competition with the goods of others, would have seemed like a basic mistake.[7] So if theorists take the central question about the origins of morality to be the evolution of altruism, they are conceptualizing the moral life in a way that has a very particular historical and social location.

Recent developments in evolutionary theory might go some way towards addressing this objection, by taking more seriously the idea that in a species as social as ours, the good of individuals is inextricably bound up with that of their neighbours.[8] Also, in recent years, attention has been paid to a wider range of moral emotions and behaviours, making the discussion less fixated on altruism.[9] Nonetheless, it remains possible for Joshua Greene to maintain that 'the essence of morality is altruism',[10] and that the moral sentiments originated as adaptations to promote group co-operation – to address the tragedy of

5 David Sloan Wilson, *Darwin's Cathedral: Evolution, Religion, and the Nature of Society* (Chicago: University of Chicago Press, 2002); see above, pp. 20–1.

6 Auguste Comte, *System of Positive Polity* (Frederic Harrison (trans.); London: Longmans, Green and Co., 1875), p. 122, online at https://openlibrary.org/works/OL761013W/System_of_positive_polity (accessed 7 November 2016).

7 Alasdair MacIntyre, *After Virtue: A Study in Moral Theory* (London: Duckworth, 2nd edn, 1985), pp. 228–29.

8 For surveys and discussion of these developments, see Celia Deane-Drummond, *The Wisdom of the Liminal: Evolution and Other Animals in Human Becoming* (Grand Rapids, MI: Eerdmans, 2014), pp. 194–237, and Markus Mühling, *Resonances: Neurobiology, Evolution and Theology. Evolutionary Niche Construction, the Ecological Brain and Relational-Narrative Theology* (Göttingen: Vandenhoeck and Ruprecht, 2014), pp. 137–64.

9 See, for example: Frans de Waal, *Good Natured: The Origins of Right and Wrong in Humans and Other Animals* (Cambridge, MA: Harvard University Press, 1996); Frans de Waal et al., *Primates and Philosophers: How Morality Evolved*, ed. Stephen Macedo and Josiah Ober (Princeton, NJ: Princeton University Press, 2006); Deane-Drummond, *The Wisdom of the Liminal*.

10 Joshua Greene, *Moral Tribes: Emotion, Reason, and the Gap between Us and Them* (London: Atlantic, 2014), p. 23.

the commons, or the problem of 'Me versus Us', faced by our evolutionary ancestors.[11]

By slipping the phrase 'the moral sentiments' into the previous paragraph, I hinted at a second noteworthy point. In much of the evolutionary literature, 'morality' is taken to be a set of intuitions or emotional predispositions leading us to make certain kinds of moral evaluations of other people and actions.[12] The judgements generated by these moral sentiments may of course then become the objects of rational reflection, but it is the existence and form of the moral sentiments that call for evolutionary explanation. This was true of early theorists like Darwin and T. H. Huxley,[13] and remains the case in the current literature. The contributors to that literature are in this sense the conceptual heirs to eighteenth-century debates involving moral-sense theorists like Shaftesbury, Hutcheson, Smith and Hume about the place of reason and the passions in moral judgement and action.[14]

These debates have also resurfaced recently in moral psychology, thanks particularly to Jonathan Haidt's dual-process model of moral cognition.[15] According to Haidt, human minds operate two different cognitive systems: a fast, effortless and automatic *intuitive system*, whose processes are not accessible to conscious awareness, and a slow, effortful *reasoning system* whose processes are intentional, consciously accessible and controllable. The typical function of moral reasoning, he maintains, is not to guide moral judgements, but to provide post hoc justifications for judgements made intuitively.[16] This

[11] Greene, *Moral Tribes*, pp. 19–65, citing Garrett Hardin, 'The Tragedy of the Commons', *Science* 168.3859 (1968), pp. 1243–48.

[12] The evolutionary – and, as we shall see, moral psychology – literature does not always distinguish as clearly between emotions and intuitions as the philosophical literature on the moral sentiments: for the latter, see Antti Kauppinen, 'Moral Sentimentalism', in Edward N. Zalta (ed.), *The Stanford Encyclopedia of Philosophy* (Spring 2014 Edition), online at http://plato.stanford.edu/archives/spr2014/entries/moral-sentimentalism/ (accessed 7 February 2016).

[13] See Charles Darwin, *The Descent of Man, and Selection in Relation to Sex* (2 vols; London: John Murray, 1871), vol. 1, pp. 158–67, online at http://darwin-online.org.uk/content/frameset?pageseq=1&itemID=F937.1&viewtype=text (accessed 7 November 2016); T. H. Huxley, 'Evolution and Ethics – Prolegomena', in *Evolution and Ethics and Other Essays* (Collected Essays, 9; London: Macmillan, 1894), pp. 1–45, online at http://aleph0.clarku.edu/huxley/CE9/index.html (accessed 7 November 2016).

[14] These connections are explored by John Perry, 'Jesus and Hume among the Neuroscientists: Haidt, Greene, and the Unwitting Return of Moral Sense Theory', *Journal of the Society of Christian Ethics* 36.1 (2016), pp. 69–85.

[15] See Jonathan Haidt, 'The Emotional Dog and Its Rational Tail: A Social Intuitionist Approach to Moral Judgment', *Psychological Review* 108.4 (2001), pp. 814–34, and Haidt, *The Righteous Mind: Why Good People Are Divided by Politics and Religion* (London: Penguin, 2012).

[16] Haidt, 'The Emotional Dog', pp. 820–22.

serves social functions, such as influencing *other* people's moral intuitions and supporting the shared convictions of groups and communities. Not everyone, however, accepts Haidt's dual-process model: others argue for a more prominent role for reason, or for a closer integration of intuitive and rational processes than Haidt allows.[17]

As already noted, these debates are in some ways a rerun of eighteenth-century arguments about the place of reason and the passions in moral judgement – sometimes explicitly, as when Haidt identifies his model with Hume's remark that '[r]eason is, and ought only to be the slave of the passions'.[18] As such, they engage with long-standing debates and may generate interesting research questions; but they focus on a particular part of the ethical landscape in a particular way, and this may mean that other parts of the picture are neglected. For example, a Kantian might say that the really important question about any moral judgement is whether it is justified, and empirical evidence about what causes people to hold that judgement is simply irrelevant.[19] As I shall suggest later, a theological survey of the territory will look rather different from either of these views.

The neuroscience of moral judgement

We come now to Greene's work itself. In this as in many neuroethical topics, the conceptualization, design and analysis of experiments is so heavily theory-laden that it is difficult to separate the scientific findings and claims from the philosophical and ethical conclusions that may be drawn from them. However, in this section, the main focus will be on Greene's experimental findings and their interpretation, while the next section will be concerned with the normative ethical arguments that Greene and others attempt to build on these findings.

[17] See, for example, David A. Pizzaro and Paul Bloom, 'The Intelligence of the Moral Intuitions: Comment on Haidt (2001)', *Psychological Review* 110.1 (2003), pp. 193–96; Michael L. Spezio, 'The Neuroscience of Emotion and Reasoning in Social Contexts: Implications for Moral Theology', *Modern Theology* 27.2 (2011), pp. 339–56. For Haidt's reply to some of these criticisms, see Jonathan Haidt, 'The Emotional Dog Does Learn New Tricks: A Reply to Pizarro and Bloom (2003)', *Psychological Review* 110.1 (2003), pp. 197–98.

[18] David Hume, *A Treatise of Human Nature* (L. A. Selby-Bigge and P. H. Nidditch (eds.); Oxford: Clarendon, 1978), section 2.3.3; Haidt, *The Righteous Mind*, p. 29.

[19] Cf. Kauppinen, 'Moral Sentimentalism', sect. 5.3.

Greene's neuroscientific studies built on a variety of earlier work, including Haidt's moral psychology and clinical studies of the effects of brain damage.[20] Greene's studies have focused on the brain activity of subjects responding to moral thought experiments related to the 'trolley problem'.[21] In his initial study, participants were asked to consider two different kinds of moral dilemma, along with non-moral dilemmas for comparison.[22] The first kind of moral dilemma, which Greene called 'impersonal', is exemplified by the following: A runaway trolley is about to collide fatally with five people on the track; would it be appropriate to save their lives by diverting it onto a siding, knowing that another person in the siding would be killed by it? The paradigm case of the second, 'personal', kind of dilemma has no siding. Instead, there is a large man (or a man wearing a heavy backpack) on a footbridge crossing the track. Would it be appropriate to stop the trolley by pushing him off the bridge into its path, sacrificing his life to save the other five? Most people answer 'Yes' to the first of these and 'No' to the second. They make a moral distinction between acting to save five lives, knowing that your action will have the consequence that one other person will die, and deliberately, directly killing one person in order to save five. A minority answer 'Yes' to both questions – a response often described in this literature as 'utilitarian', which *apparently* denies any morally relevant difference between the two scenarios.[23]

Greene et al. asked their subjects to consider a range of dilemmas classified as non-moral, moral-personal and moral-impersonal, and assessed the activity of various brain regions using fMRI.[24] They found that moral-personal

20 On connections between Greene's and Haidt's work, see Joshua Greene and Jonathan Haidt, 'How (and Where) Does Moral Judgment Work?', *Trends in Cognitive Sciences* 6.12 (2002), pp. 517–23. The relevant clinical studies concern patients with damage to the ventromedial prefrontal cortex (VMPFC), an area involved in emotional processing, who exhibit impaired social functioning and an inability to make prudent decisions: see Antonio Damasio, *Descartes' Error: Emotion, Reason and the Human Brain* (New York: Putnam, 1994). The classic case history of VMPFC damage is that of the nineteenth-century railway construction foreman Phineas Gage, retold in *Descartes' Error*, ch. 1.

21 Since its invention by Philippa Foot the trolley problem has spawned numerous variants and extensive philosophical discussion, sometimes of quite baroque complexity. For a recent treatment see F. M. Kamm with Judith Jarvis Thomson, Thomas Hurka, and Shelley Kagan, *The Trolley Problem Mysteries* (Eric Rakowski (ed.); Oxford: Oxford University Press, 2015).

22 Joshua D. Greene et al., 'An fMRI Investigation of Emotional Engagement in Moral Judgment', *Science* 293 (2001), pp. 2105–8.

23 As we shall see, however, this description of the minority response is disputed by some commentators.

24 Greene et al., 'An fMRI Investigation of Emotional Engagement'. The moral dilemmas were classified as 'personal' if and only if the action in question '(a) could reasonably be expected to lead to serious bodily harm (b) to a particular person or a member or members of a particular group of people (c) where this harm is not the result of deflecting an existing threat onto a different party' (pp. 2107–8 n. 9). Participants had to be given numerous different scenarios in each of the three categories in order to generate a statistically significant number of responses. This did however give rise to a problem discussed below.

scenarios were associated with greater activity in areas of the brain implicated in emotional processing,[25] whereas moral-impersonal and non-moral scenarios were associated with higher activity in areas involved in cognitive processing.[26] They also measured reaction times, which were significantly longer for participants approving 'utilitarian' actions in the moral-personal scenarios (for example, pushing the man off the footbridge) than for those condemning those actions. There was no significant difference in reaction times for different responses to the non-moral or moral-impersonal dilemmas. Greene et al. inferred that in the moral-personal cases, the 'utilitarians' were using their reason to override their non-utilitarian intuitive judgements.

According to Greene, these findings support a dual-process model of moral cognition similar to Haidt's, in which subjects' different responses in the fMRI study arise from interactions between two different subsystems in the brain. One subsystem involves areas implicated in emotion and social cognition, and generates rapid, domain-specific, intuitive judgements. The other, involving areas associated with abstract reasoning and problem-solving, works more slowly to generate reasoned conclusions.[27] More recently, Greene has used the metaphor of a digital camera with two modes, likening the rapid intuitive process to an automatic 'point-and-shoot' mode, and the reasoning process to a slower manual mode.[28] On the strength of subsequent studies, he has argued more specifically – and controversially – that the 'manual mode' generates utilitarian judgements whereas deontological judgements arise from the 'automatic mode'.[29] Studies by other researchers have indicated higher levels of utilitarian moral judgement in patients with ventromedial pre-frontal cortex (VMPFC) damage and in individuals with one subtype of psychopathy.[30]

[25] Specifically, the medial frontal gyrus, posterior cingulate gyrus, and bilateral superior temporal sulcus.

[26] The dorsolateral prefrontal cortex and parietal lobes; see the summary in Greene and Haidt, 'How (and Where) Does Moral Judgment Work?', p. 519.

[27] Ibid.; Joshua D. Greene et al., 'The Neural Bases of Cognitive Conflict and Control in Moral Judgment', *Neuron* 44 (2004), pp. 389–400 (389–90). However, Greene holds that his dual-process model gives a greater place to reason than Haidt's.

[28] Greene, *Moral Tribes*, pp. 132–43 and passim.

[29] Greene et al., 'Neural Bases of Cognitive Conflict'; Joshua D. Greene et al., 'Cognitive Load Selectively Interferes with Utilitarian Moral Judgment', *Cognition* 107 (2008), pp. 1144–54.

[30] Michael Koenigs et al., 'Damage to the Prefrontal Cortex Increases Utilitarian Moral Judgements', *Nature* 446 (2007), pp. 908–11; Michael Koenigs et al., 'Utilitarian Moral Judgment in Psychopathy', *Social, Cognitive and Affective Neuroscience* 7 (2012), pp. 708–14. For a summary of evidence supporting Greene's dual-process theory, see Joshua D. Greene, 'Beyond Point-and-Shoot Morality: Why Cognitive (Neuro)Science Matters for Ethics', *Ethics* 124.4 (2014), pp. 695–726 (700–6).

Greene's research has been criticized in various ways for its conceptualization, design and interpretation. First, the scenarios given to participants in Greene's initial study were quite varied, and there could have been significant differences between those grouped together in each category. This may have distorted the results and interpretation, undermining the distinction between 'personal' and 'impersonal' scenarios, and questioning the dual-process theory that Greene builds on the results.[31] Greene has acknowledged the problem with the personal/impersonal distinction,[32] but maintains that the dual-process theory is nonetheless supported by a range of other evidence.[33]

Second, the extensive use of trolley-like scenarios has been criticized on various grounds, for example that they are far-fetched, sometimes highly intricate, and at best give an incomplete and one-sided picture of the psychology of moral agents in the real world. For example, Guy Kahane argues that focusing mostly on these scenarios creates a spurious association of reason with utilitarian, and intuition with deontological, judgements.[34] He and his colleagues report that when subjects are asked to consider other types of moral dilemma, the distinction between intuitive and reasoned responses maps onto the intuitive or counterintuitive character of the moral judgement made, not whether it is utilitarian or deontological.[35] Greene and his colleagues have presented counter-evidence against Kahane's claim,[36] but this dispute remains unresolved at the time of writing.

[31] Jonathan McGuire et al., 'A Reanalysis of the Personal/Impersonal Distinction in Moral Psychology Research', *Journal of Experimental Social Psychology* 45 (2009), pp. 577–80.

[32] He now argues that the moral brain's 'automatic mode' is sensitive both to the use of personal force and to the distinction between means and side-effects (roughly speaking, the distinction articulated in the doctrine of double effect); intuitive non-utilitarian judgements, he maintains, are evoked particularly strongly by scenarios that *combine* personal force with the use of harm as a means: see, for example, Greene, *Moral Tribes*, pp. 211–25.

[33] Joshua D. Greene, 'Dual-process Morality and the Personal/Impersonal Distinction: A Reply to McGuire, Langdon, Coltheart, and Mackenzie', *Journal of Experimental Social Psychology* 45 (2009), pp. 581–84.

[34] Guy Kahane, 'On the Wrong Track: Process and Content in Moral Psychology', *Mind and Language* 27.5 (2012), pp. 519–45; idem, 'Sidetracked by Trolleys: Why Sacrificial Moral Dilemmas tell us little (or nothing) about Utilitarian Judgment', *Social Neuroscience* 10.5 (2015), pp. 551–60.

[35] Guy Kahane et al., 'The Neural Basis of Intuitive and Counterintuitive Moral Judgment', *Social, Cognitive and Affective Neuroscience* 7 (2012), pp. 393–402. For example, in a scenario which raises the question, 'is it right to kill one person in order to save five?', it is the 'utilitarian' answer that appears counter-intuitive and requires more effortful reasoning; but when the question is, 'is it right to tell a lie to save someone from emotional harm?', it is the 'deontological' answer.

[36] Joseph M. Paxton, Tommaso Bruni and Joshua D. Greene, 'Are "Counter-Intuitive" Deontological Judgments Really Counter-Intuitive? An Empirical Reply to Kahane et al. (2012)', *Social, Cognitive and Affective Neuroscience* 9 (2014), pp. 1368–71.

Furthermore, Greene's representation of 'utilitarian' and 'deontological' judgements is (according to some of his critics) simplistic and misleading. Kahane maintains that trolley-like dilemmas do not elicit distinctively utilitarian judgements in which the sole criterion of rightness would be to maximize welfare: the decision to push the man off the footbridge to stop the trolley killing five others, for example, could be motivated by a deontological rule to prevent harm.[37] If so, the cognitive effort involved in the decision would be needed to resolve the conflict between two *prima facie* duties generated by that principle. According to Kahane, results like Greene's are better understood in terms of a 'commonsense morality' which is diverse, untidy, pluralist, and *non*-utilitarian.[38] In fairness to Greene, however, he need not claim that those making 'utilitarian' judgements in empirical studies have anything like a fully worked-out utilitarian theory in their minds that they apply in a consistent manner to the scenarios. It seems closer to his view to say that the more subjects reason about their judgements, the more they are performing the kind of benefit-maximizing calculation that Bentham and Mill systematized in utilitarian ethical theory. Whether even this claim is supported by the evidence, of course, is a matter of dispute between Greene and his critics.

What should a theologian listening in to these debates conclude? Like so much in neuroethics, the neuroscience of morality is a new research area whose conclusions are still tentative, in part because of the current limitations of imaging technologies and because the functions of some brain areas implicated in moral judgement are not well understood.[39] Moreover, the early development of this area may have been unbalanced by the focus and design of Greene's work, as critics charge. Nevertheless, some provisional conclusions may be drawn. First, this work suggests that emotion plays a large part in moral judgement, and much moral decision-making is done by automatic rather than consciously controlled cognitive processes. Although Kahane thinks this will not come as a surprise to philosophers,[40] it has often been surprisingly

[37] Kahane, 'On the Wrong Track'; idem, 'Sidetracked by Trolleys'. See also F. M. Kamm, 'Neuroscience and Moral Reasoning: A Note on Recent Research', *Philosophy and Public Affairs* 37.4 (2009), pp. 330–45.

[38] Kahane, 'Sidetracked by Trolleys', pp. 554–55.

[39] For one example see Greene et al., 'Neural Bases of Cognitive Conflict', p. 395.

[40] Kahane, 'On the Wrong Track', p. 519.

easy to forget.[41] The work of Haidt, Greene and others could therefore serve as a useful reminder, at the very least.

Second, it lends some support to some kind of dual-process theory of moral judgement.[42] Even Kahane, one of Greene's more stringent critics, allows this, though he does not think the dual-process theory that survives his criticisms is either surprising or interesting: 'Intuitive judgments are generally associated with automatic processing, and counterintuitive judgments with controlled processing.'[43] It might however be possible to go further, and conclude that at least some of what we experience as reasoned moral decision-making is in fact post hoc rationalization of intuitive judgements. Again, this might be something we should have known all along, but if so it has often proved easy to forget, and in any case empirical work that offers greater understanding of these processes could perform a useful critical function. The nature and extent of the parts played by automatic and controlled processes, of course, is less clear as yet.

Third, Greene's claim that deontological judgements are generated by intuitive processes, and utilitarian judgements by controlled ('manual mode') processing, is less well established. Some of the criticisms levelled at his experimental design, particularly his reliance on trolley-like problems, raise questions about it. He could be right, of course, but the jury is still out.

Finally, there is a broader concern about ecological validity. Not only Greene's neuroimaging experiments, but many psychological studies, are based on artificial and sometimes rather fantastical moral thought-experiments. Some artificiality may be inevitable in order to generate results that can be systematically analysed. However, philosophers like Alasdair MacIntyre and theologians like Stanley Hauerwas have long complained that 'quandary ethics' offers a seriously inadequate picture of the moral life, which is better understood as an embodied, socially connected narrative than a series of

[41] For example, as Haidt points out, during the rationalist phase of moral psychology in the 1960s and 1970s: 'The Emotional Dog', pp. 816–17.

[42] So also Selim Berker, 'The Normative Insignificance of Neuroscience', *Philosophy and Public Affairs* 37.4 (2009), pp. 293–329 (313).

[43] Kahane, 'On the Wrong Track', p. 540. Others, however, contest the case for any kind of dual-process theory. Michael Spezio, for example, argues that such theories take insufficient account of the evidence from clinical neurology, which is better accounted for by an 'integrationist' model in which emotion plays a constitutive role in reasoning: Spezio, 'The Neuroscience of Emotion and Reasoning'.

dilemmas to be solved.[44] The gulf between a moral life so understood and the performance of artificial and sometimes far-fetched thought experiments in a laboratory setting is evident. That gap becomes all the wider when the laboratory setting is the alien environment of an fMRI scanner, as Uffe Schjoedt has also observed about the neuroscience of religion.[45] To my mind that is not a reason to dismiss this research, but it does suggest that caution is needed in drawing conclusions about the real moral lives of human beings and communities from the findings.[46] It should also, of course, motivate the search for experimental approaches which narrow the gap.[47]

In this section, the focus has been on descriptive and explanatory questions: what is going on in our brains when we make moral judgements, and what that tells us about how those judgements are made. It should be clear from my account how heavily theory-laden, and how contested, even the descriptive and explanatory conclusions are. Their use to support normative ethical arguments, however, is a further step – and by no means a straightforward one. In the next section, we turn our attention to that further step.

Normative ethical implications

Both Greene and Peter Singer have argued that cognitive-science research on morality indirectly supports utilitarianism, by indicating that deontologists' reasons for rejecting distinctively utilitarian judgements are based on morally irrelevant factors.

[44] MacIntyre, *After Virtue*; Stanley Hauerwas with Richard Bondi and David B. Burrell, *Truthfulness and Tragedy: Further Investigations into Christian Ethics* (Notre Dame, IN: University of Notre Dame Press, 1977), pp. 15–39.

[45] See Uffe Schjoedt, 'The Religious Brain: A General Introduction to the Experimental Neuroscience of Religion', *Method and Theory in the Study of Religion* 21 (2009), pp. 310–39 (314).

[46] More negative assessments are given by Spezio, 'The Neuroscience of Emotion and Reasoning', and Harris Wiseman, *The Myth of the Moral Brain: The Limits of Moral Enhancement* (Cambridge, MA: MIT Press, 2016), ch. 5. Wiseman argues that such studies are worse than useless, because they tell us nothing meaningful about morality but give us the illusion of understanding something about it. However, it is not clear to me why an acknowledgement of the problems with Greene's and others' studies should drive us to this conclusion rather than suggesting (as I would say) that they are problematic in various ways and should be handled with a good deal of caution – but used with such caution, may suggest interesting insights and questions.

[47] An alternative research programme to Greene's, focusing on the virtues and moral exemplars, is set out in James A. Van Slyke et al. (eds.), *Theology and the Science of Moral Action: Virtue Ethics, Exemplarity, and Cognitive Neuroscience* (New York: Routledge, 2013).

Singer's version of this debunking argument can be summarized, briefly, as follows.[48] Utilitarianism notoriously generates conclusions that many people intuitively find repugnant, as in the footbridge scenario and others like it. A standard critique is that there must be something wrong with the theory if it gives these repugnant results; the question is whether such moral intuitions indicate insights that should be taken seriously. Some ethical approaches, such as John Rawls's method of reflective equilibrium, suggest that they may.[49] Singer argues that such approaches are undermined by scientific studies. Evolutionary accounts suggest our moral intuitions evolved to promote co-operation in (relatively small) social groups. We could therefore be expected to have strong intuitive reactions to situations involving up-close-and-personal harmful actions, but not those where harm is caused remotely, simply because the up-close-and-personal sort were the only ones our evolutionary ancestors encountered. They did not have the technology to cause actions at a distance, as in the trolley scenario. According to Singer, this prediction is borne out by Greene's data. But, he maintains, 'the fact that I have killed someone in a way that was possible a million years ago, rather than in a way that became possible only two hundred years ago' has no moral salience.[50]

Greene builds on this kind of debunking argument, incorporating it into a large-scale argument for utilitarianism.[51] As we have seen, he likens the two processes of our moral psychology to modes of a digital camera: intuitive morality to an 'automatic mode', and moral reasoning to a more flexible but slower 'manual mode'. The automatic mode has evolutionary origins. It is an effective heuristic for producing adaptive solutions to the problem of 'Me versus Us': averting the tragedy of the commons that would result from group members putting their individual interests ahead of the group's claims. But nowadays a different problem arises: a modern society is a mix of people belonging to diverse communities with different group norms and shared convictions (different 'moral tribes', in Greene's parlance), who

[48] Peter Singer, 'Ethics and Intuitions', *Journal of Ethics* 9 (2005), pp. 331–52.

[49] See John Rawls, *A Theory of Justice* (Cambridge, MA: Harvard University Press, 1971); Norman Daniels, 'Reflective Equilibrium', in *The Stanford Encyclopedia of Philosophy* (ed.) Edward N. Zalta (Winter 2013), online at http://plato.stanford.edu/archives/win2013/entries/reflective-equilibrium/ (accessed March 12, 2015).

[50] Singer, 'Ethics and Intuitions', p. 348.

[51] Greene, *Moral Tribes*, esp. part 2, 'Morality Fast and Slow' (pp. 103–44) and part 3, 'Common Currency' (pp. 145–208).

have to find ways of getting along together. This creates a new tragedy, the 'Tragedy of Commonsense Morality' (or the problem of 'Us versus Them'), because different groups have different, sometimes incompatible, sets of group norms, so their automatic moral intuitions become a source of intractable conflict.[52] Intuitive moral judgements cannot resolve these conflicts, so moral reasoning is required. However – this is where the debunking argument comes in – empirical moral psychology and neuroscience show that much of our moral reasoning simply constructs post hoc rationalizations for our intuitive moral judgements. Greene makes this claim about deontological reasoning in particular, following Singer's criticism of Rawls but also critiquing rights language and Kantian theory.[53] The kind of moral reasoning Greene thinks we need is a 'metamorality', a mode of reasoning that will transcend our tribe-specific instincts. He believes this is supplied by utilitarianism, which offers a 'common currency' for settling our differences by means of empirically testable proposals about the best outcomes for all concerned. It is worth emphasizing that it is non-utilitarian moral *theory* that Greene wishes to reject, not our intuitive judgements themselves. He believes they perform better than 'manual-mode' reasoning for many of our everyday 'Me versus Us' moral problems, because our moral reasoning tends to be unconsciously biased in our own favour. But when we encounter moral controversy, that is a sign of an 'Us versus Them' problem, and then (utilitarian) reasoning is required.[54]

Various critical remarks may be made about Singer's and Greene's arguments. First, Singer's debunking argument depends heavily on the distinction between personal and impersonal dilemmas in the first neuroimaging study by Greene et al. As we have seen, Greene later acknowledged the distinction in this form to be flawed, which on the face of it casts doubt on Singer's argument. However, this is probably not a serious objection. Greene's more recent proposal, that

[52] Readers who think they hear echoes of MacIntyre's observation about the fragmentation and incoherence of modern moral discourse in the first chapter of *After Virtue* might not be mistaken: see Greene, *Moral Tribes*, pp. 381–82. I shall return to this point in the final section.

[53] Regarding Kant, Greene makes much of a Nietzschean aphorism mocking Kant's argument against masturbation, which (he maintains) exposes Kant's argument as a rationalization of the repressed sexual mores of his time and place. He generalizes from this and other examples to the rather bold conclusion that Kant's entire ethical theory is an exercise in post hoc rationalization: Joshua D. Greene, 'The Secret Joke of Kant's Soul', in Walter Sinnott-Armstrong (ed.), *Moral Psychology, Vol. 3: The Neuroscience of Morality: Emotion, Brain Disorders, and Development* (Cambridge, MA: MIT Press, 2008), pp. 36–79; *Moral Tribes*, pp. 301, 331, 332.

[54] Greene, *Moral Tribes*, pp. 293–98.

strong emotional responses are evoked by scenarios that combine the personal nature of the harm with its being a means rather than a side-effect, would accomplish similar argumentative work in Singer's argument.[55]

This, however, relates to a second point. Singer simply ignores another distinction between the trolley and footbridge scenarios: in the first, the death of the person on the siding is a foreseen consequence of saving the five on the main line, but need not be intended; in the second the death of the person pushed off the bridge is the *means* by which the five are saved.[56] Of course Singer, as a utilitarian, presumably does not consider this distinction morally relevant. However, he cannot simply take its irrelevance for granted if he hopes to convince non-utilitarians, since at least some of them will consider the distinction between means and unintended consequences highly relevant.

Greene does not ignore this distinction, but he does dismiss it. His grounds for doing so are that those who defend it appeal to the doctrine of double effect, but the latter is no more than an attempt to systematize moral intuitions about specific cases.[57] If Greene succeeds in debunking the doctrine of double effect and the distinction between means and unintended consequences, this will be a telling point in his favour, given their importance in a range of real-world problems.[58] Unfortunately, he does not make his case very convincingly. First, he gives the impression that everyone who defends the means/unintended consequences distinction appeals to double effect, which is not the case.[59] More importantly, he does not offer much in the way of reasons for debunking the double effect doctrine. In a recent article, he claims supporting evidence from 'the willingness of philosophers to abandon [the doctrine] when it fails to get the intuitively right answers', and cites an experimental study comparing the

[55] Ibid., pp. 217–24.

[56] Some elaborations of trolley-like problems complicate this distinction. For example, in one of these the single person is on a loop of track, and if the trolley does not collide with this person it will return to the main line and kill the other five: Judith Jarvis Thomson, 'The Trolley Problem', *Yale Law Journal* 94.6 (1985), pp. 1395–1415 (1402–3). These cases certainly complicate the picture, though perhaps they say more about the ingenuity of analytic moral philosophers than about the evaluation of acts in a real human life; in any event, they do not undermine my basic point, that there are factors not considered by Singer which many ethicists (and others) would consider relevant.

[57] Greene, *Moral Tribes*, pp. 223–24.

[58] Ibid., pp. 252–53.

[59] For a brief survey see Alison McIntyre, 'Doctrine of Double Effect', *The Stanford Encyclopedia of Philosophy* (Winter 2014 Edition), Edward N. Zalta (ed.), online at http://plato.stanford.edu/archives/win2014/entries/double-effect (accessed 15 February 2016).

moral judgements of philosophers and non-philosophers.[60] But this by itself does not amount to a very strong case for its non-credibility. He may be right that it can be debunked, but he has not yet shown this convincingly.[61]

A third critical point is Selim Berker's 'worry' that neuroscience actually does no argumentative work in Singer's and Greene's arguments. What is doing the work, he thinks, is that 'from the armchair' Greene (1) identifies dilemmas eliciting deontological responses as personal and those eliciting consequentialist responses as 'impersonal', then (2) claims that the personal/impersonal distinction is morally irrelevant.[62] Greene's response is that empirical studies do real work in the normative argument, in both direct and indirect ways. The direct way is that they can provide factual information which, in combination with normative premises, can yield new and interesting normative insights.[63] The indirect way is that they suggest the kinds of dilemmas for which our moral intuitions are likely to prove unreliable: broadly speaking, those whose answers have not been shaped by trial and error through natural selection, cultural transmission or individual learning.[64] His own summary of the contribution of cognitive science to ethics is that it can '[reveal] the hidden inner workings of our moral judgments, especially the ones we make intuitively. Once those inner workings are revealed we may have less confidence in some of our judgments and the ethical theories that are … based on them'.[65] This has some similarity to Gregory Peterson's proposal that the cognitive science of religion can provide a hermeneutic of suspicion regarding religious experiences and beliefs.[66] As I shall suggest later, this could make Greene an unexpected ally of theological ethics.

[60] Greene, 'Beyond Point-and-Shoot Morality', p. 721.

[61] However, the theological analysis in the following sections will give reasons to think that there is less at stake in this for theological ethics than might be supposed.

[62] Berker, 'The Normative Insignificance of Neuroscience', pp. 325–26.

[63] Greene, 'Beyond Point-and-Shoot Morality', pp. 711–13.

[64] Ibid., pp. 713–17. The individual learning he has in mind is the kind that happens when we burn ourselves on hot objects, not the kind that goes on in philosophy classes.

[65] Ibid., pp. 695–96. There are echoes here of arguments about the significance of scientific explanations of religion, discussed in Chapter 2. A critic might object that a causal explanation of the origins of our moral judgements is simply irrelevant to their normative force; what matters is whether there are good *reasons* for holding them. Greene could perhaps reply that causal explanations do not refute deontological arguments, but do show them to be epistemically dubious.

[66] Gregory R. Peterson, 'Are Evolutionary/Cognitive Theories of Religion Relevant for Philosophy of Religion?' *Zygon* 45.3 (2010), pp. 545–57.

Finally, in a detailed critique of evolutionary debunking arguments, Guy Kahane has argued that they might turn out to debunk more than Singer and Greene want them to.[67] He cites recent 'global' evolutionary debunking arguments in metaethics,[68] but argues that even if these prove incorrect, Singer and Greene cannot be confident that the debunking will stop short of utilitarianism. What is left standing might resemble Nietzsche more than Bentham or Mill.[69] Greene might be able to rescue his defence of utilitarianism from this threat if he is right that consequentialist judgements are generated by a different cognitive system, with a less domain-specific function, from deontological judgements.[70] However, *evolutionary* debunking is not the only kind, and here Nietzsche might prove a dangerous ally, since his criticisms of utilitarianism were just as excoriating as those he directed at Kant.[71]

To sum up regarding Greene's cognitive neuroscience-based defence of utilitarianism: his debunking of some non-utilitarian judgements has a certain plausibility, though in other cases – for example double effect – he does not offer a very persuasive case. He *might* be able to stop the evolutionary debunking trolley (so to say) before it flattens utilitarianism as well as deontology; that may depend on whether his empirical claim, that utilitarian and deontological judgements are produced by different cognitive processes, proves robust. Even if he can, other forms of debunking could still undermine the role he wishes to claim for utilitarianism, as a 'metamorality' that transcends the differences between 'moral tribes'.

Theological suspicion

Earlier I suggested that the conceptual and scientific criticisms of Greene's research give theological readers reason to receive his findings cautiously and critically, not to dismiss or ignore them. Assuming for the sake of argument

[67] Guy Kahane, 'Evolutionary Debunking Arguments', *Noûs* 45.1 (2011), pp. 103–25.

[68] For example, Richard Joyce, *The Evolution of Morality* (Cambridge, MA: MIT Press, 2006) and Sharon Street, 'A Darwinian Dilemma for Realist Theories of Value', *Philosophical Studies* 127 (2006), pp. 109–66.

[69] Kahane, 'Evolutionary Debunking Arguments', p. 120.

[70] Cf. Greene, 'The Secret Joke', p. 64.

[71] See Jonny Anomaly, 'Nietzsche's Critique of Utilitarianism', *Journal of Nietzsche Studies* 29 (2005), pp. 1–15.

that such studies, if cautiously handled, may yield genuine insights and questions, how should a Christian ethic engage with the insights and address the questions?

In the remainder of the chapter, I shall set out one proposal for such a theological engagement.[72] In this section, I shall argue that Greene's critical perspective on intuitive moral judgements may make him an unexpected ally for the theological suspicion of ethics voiced powerfully by Karl Barth and Dietrich Bonhoeffer. In the next, the focus shifts to the theological reconstruction of ethics that follows this theological critique. The final section outlines some more concrete implications of the kind of theological response I have proposed.

In the Genesis 'Fall' narrative, the serpent promises that if they eat the prohibited fruit, the human beings 'will be like God, knowing good and evil' (Gen. 3.5). Commenting on this text, Barth acidly remarks, 'What the serpent has in mind is the establishment of ethics'.[73] In similar vein, one fragment of Bonhoeffer's unfinished *Ethics* begins:

> The knowledge of good and evil appears to be the goal of all ethical reflection. The first task of Christian ethics is to [overcome] that knowledge. This attack on the presuppositions of all other ethics is so unique that it is questionable whether it even makes sense to speak of a Christian ethics at all. If it is nevertheless done, then this can only mean that Christian ethics claims to articulate the origin of the whole ethical enterprise, and thus to be considered an ethic only as the critique of all ethics.[74]

Here Bonhoeffer is hinting at a different kind of narrative of the origins of ethics from the scientific origin-narratives that we have been discussing in this chapter: a theological narrative of origins of the sort that he earlier expounded

[72] A different kind of cautiously positive reception of Haidt and Greene is offered by Perry, 'Jesus and Hume among the Neuroscientists'; but for a more negative assessment of what they might have to offer theological ethics, see Spezio, 'The Neuroscience of Emotion and Reasoning'.

[73] Karl Barth, *Church Dogmatics* (Geoffrey W. Bromiley and Thomas F. Torrance (ed.), 13 vols, Edinburgh: T&T Clark, 1956–75), vol. 4.1, p. 448.

[74] Dietrich Bonhoeffer, *Ethics* (Ilse Tödt et al. (ed.); Reinhard Krauss, Charles C. West and Douglas W. Stott (trans.); Dietrich Bonhoeffer Works, 6; Minneapolis, MN: Fortress, 2005), pp. 299–300. In the second sentence of the quotation, I have amended 'supersede' in the published translation to read 'overcome', which to my mind gives a better sense of the German '*aufheben*' and is indeed routinely used by the translators in the remainder of the fragment (pp. 311ff.).

in *Creation and Fall*, his 1932–33 Berlin lectures on Genesis 1–3.[75] This is not, of course, some kind of creationist refusal of the scientific narrative, as Bonhoeffer makes clear elsewhere (pp. 62, 76). It is more akin to Barth's reading of the Genesis texts as 'sagas', or narratives disclosing the divine origins of creation that lie beyond the reach of scientific investigation.[76] The theological origin-narrative is not a rival to scientific narratives, but can perfectly well be read alongside them.

In *Creation and Fall*, Bonhoeffer argues that God's prohibition against eating from the tree of the knowledge of good and evil (Gen. 2.17) is grace. The limit set by God makes possible Adam's creaturely freedom and knowledge of God: 'Adam knows neither what good nor what evil is and lives in the strictest sense *beyond good and evil*; that is, Adam lives out of the life that comes from God, before whom a life lived in good, just like a life lived in evil, would mean an unthinkable falling away' (pp. 86–87, emphasis original).[77] The humans' disobedience, incited by the serpent, is a refusal of that creaturely limit (and therefore of their creaturehood as such) in the attempt to become *sicut deus*, 'like God'. It is a successful attempt; in that respect at least, the serpent is not wrong: 'the fall *really* makes the creature – humankind in the imago dei – into a creator sicut deus', writes Bonhoeffer (p. 116, emphasis original). But this is a radically alienating development. The unity of knowing God as Creator, 'as the center and the boundary of human life' (p. 87), is broken apart into the knowledge of good and evil (*tob* and *ra*). The refusal of the creaturely limit leaves no room in the human being's universe for his or her Creator, so when God says to Adam, 'where are you?' Adam's response is to run away and hide. As beings separated from God by the division of our knowledge into good and evil, we are also divided within ourselves; and this dividedness from ourselves and flight from our Creator, remarks Bonhoeffer, 'we call conscience' (p. 118). The knowledge of *tob* and *ra*, he argues, also gives rise to shame and obsessive desire, estranging us from one another; and our refusal to live as human creatures divides or estranges us from nature, from the 'soil' and from our work in the world.

[75] Dietrich Bonhoeffer, *Creation and Fall: A Theological Exposition of Genesis 1–3* (Martin Rüter, Ilse Tödt and John W. de Gruchy (ed.); Douglas Stephen Bax (trans.); Dietrich Bonhoeffer Works, 4; Minneapolis, MN: Fortress, 2004). Page references in brackets in the text in the following paragraphs are to this work.

[76] Barth, *Church Dogmatics*, vol. 3.1, pp. 81–94.

[77] The allusion to Nietzsche is hardly a coincidence: there is an ongoing critical dialogue with him throughout this part of *Creation and Fall*. Bonhoeffer appears to be using Nietzsche somewhat similarly to the way MacIntyre later does in *After Virtue*: as a penetrating critic of the rule-based moralities that he, Bonhoeffer, aims to 'overcome'.

This reading of the Fall narrative underpins the profound theological suspicion of ethics-as-human-project that Bonhoeffer articulates in *Ethics* when he writes that the first task of Christian ethics is to overcome the knowledge of good and evil. Ethics, understood as the knowledge of good and evil, is always already part of humanity's fallen condition, our 'disunion and estrangement' from the 'origin' in which human beings 'know nothing but God alone'.[78] It is an aspect of the life we are compelled to live in the world, estranged from God, out of our own resources; a life lived, as Bonhoeffer says, 'between curse and promise'.[79]

If Bonhoeffer is right that Christian theology should understand the human project of ethics in this way, then the critical perspective supplied by Greene's neuroscientific studies should be remarkably untroubling to Christians. Indeed, as I suggested earlier, scientists who offer such critiques might even prove to be unexpected allies (perhaps despite themselves) of Christian ethics. First, it should come as no particular shock if Haidt and others are correct that many of what we take to be reasoned moral judgements are in fact post hoc rationalizations of affect-driven moral intuitions. If this research suggests that we humans have a deep-seated capacity for self-deception about what is actually going on in our moral psychology, and a falsely inflated sense of the power and reliability of our practical reason, this looks like something Christian ethics should already know very well about the fallen human condition. If it comes as news to us, perhaps cognitive neuroscience is giving us a useful reminder.

On a larger scale, the picture of ethics offered by Greene and others is that our intuitive knowledge of good and evil has evolutionary origins as an adaptation for life in social relationships – addressing the problem of 'Me versus Us'. We might express this theologically, following Bonhoeffer, by describing our moral intuitions as means by which we seek to live in the world from our own resources (which we are compelled to do but cannot), endeavouring to overcome our estrangement from one another. Yet according to the scientific accounts surveyed earlier in the chapter, the intuitive morality that fosters in-group cohesion also promotes out-group hostility. In Greene's parlance, it addresses the problem of 'Me versus Us', but at the cost of creating the problem of 'Us versus Them', of 'moral tribes'. In other words (in Bonhoefferian terms),

[78] Bonhoeffer, *Ethics*, p. 300. As an editorial note explains, *Entzweiung*, variously translated 'disunion', 'estrangement', 'division', etc., is a key category in this fragment of *Ethics* (again echoing *Creation and Fall*, e.g., p. 88).

[79] Bonhoeffer, *Creation and Fall*, p. 135.

our efforts to overcome our disunion through our knowledge of good and evil merely replicate that disunion. As I shall argue in the next section, however, Greene's proposed solution – the use of utilitarian reason to transcend the differences between our 'tribal' moralities – will be subject to the same critique. His proposal for overcoming our disunion also replicates that disunion.

Greene might reply that Christian morality is every bit as 'tribal' as any other, and therefore it too is part of the problem rather than the solution. In a sense, a Christian theologian who takes my proposed view need not disagree. This theological critique is just as much a critique of *Christian* ethics as any other, to the extent that Christian ethics becomes disconnected from its origin and turns into another human effort to grasp the knowledge of *tob* and *ra*. And although some claims about the divisiveness of Christianity might be distorted, misleading or ideologically driven, we cannot deny that Christian communities too often make their own contribution to the problem of 'Us versus Them'. The Church too stands under judgement – though, to repeat, theologians who acknowledge this problem will disagree with Greene about the remedy.

In short, Barth and Bonhoeffer articulate a profound theological suspicion of ethics-as-human-project: the attempt to know about the good on the strength of our own resources of reason and insight. If the critical perspectives on moral judgement offered by Haidt, Singer and Greene are broadly correct, then these authors are playing the helpful ground-clearing role of 'masters of suspicion': unmasking the pretentions of a merely human project of ethics and recalling theological ethics to its proper sources and methods.[80]

Theological reconstruction

As Bonhoeffer makes plain, however, theological suspicion does not mean the wholesale rejection of ethics, but rather its transformation and reconstruction

[80] It might be asked what are the consequences for my account if Haidt and Greene turn out *not* to be correct. In my view, the consequences will not be particularly serious, because the theological suspicion of ethics-as-human-project I propose is developed on theological, not scientific, grounds. The findings of Greene and others, if correct, simply serve as a potentially helpful reminder of that theologically-motivated suspicion. The point is to show first that Greene, so far from threatening a theological ethic of this sort, may prove an ally to it, and second, that unless one accepts his contestable philosophical presuppositions, his findings may lead somewhere quite different from the conclusions that he thinks they support.

in the light of the Gospel. The 'knowledge of good and evil', as we have seen, he associates with the Fall and the curse that comes in its wake: fallen humanity cannot live out of its own resources yet is compelled to do so. 'That', remarks Bonhoeffer in *Creation and Fall*, 'is what death means' (p. 135). And yet the curse itself contains concealed within it God's compassion and promise:

> [N]ow death becomes for human beings, who live because they are preserved in compassion, a promise held out to them by the God of grace ... How should Adam know that, in this promise of death, already the end of death, the resurrection of the dead, was being spoken of? (pp. 135–36)

The story does not end with the Fall. Even the narrative of the Fall points ahead to God's promise of reconciliation in Jesus Christ.

In one fragment of the *Ethics*, Bonhoeffer takes up this theme, remarking that '[t]o anyone who reads the New Testament even superficially, it must be apparent that here this entire world of disunity, conflict and ethical problems seems to have vanished out of sight'.[81] With reference to gospel narratives of Jesus' sayings and encounters with others, he contrasts the old state of disunion and estrangement with the new situation of 'rediscovered unity [and] reconciliation' (p. 309) depicted in the New Testament.[82] Life in the state of disunion is intensely focused on conflicts and questions about good and evil. Indeed, it is the most morally serious and admirable people whose 'thoughts day and night are intensely focused on the unfathomable number of possible conflicts in order to think them through in advance, come to a decision, and determine their own choice' (p. 310). *Judging* between good and evil is at the heart of the moral life in this state of disunion; human beings in this state are 'essentially judges', and as such are 'like God, but with the difference that each judgment they pass falls on themselves' (p. 314). *Self-knowledge* is another feature of this life: humans in this state are self-consciously aware of the good they have done – and, naturally, grateful to God for it (p. 317; cf. Lk. 18.9-14).

[81] Bonhoeffer, *Ethics*, pp. 299–338 (309). Page references in the text in the following paragraphs are to this work.

[82] Bonhoeffer sets up this contrast of the old and new in terms of Jesus' encounters with the Pharisees, but goes on to use the term 'Pharisees' generically to denote 'human beings for whom nothing but the knowledge of good and evil has come to be important for their entire lives ... the epitome of [human beings] in the state of disunion' (pp. 309–10). This generic use of the label 'Pharisee' is much more problematic in our time than it might have seemed to Bonhoeffer, and will be avoided in the following account.

According to Bonhoeffer, in the Gospel narratives, Jesus simply refuses this way of life. When either his friends or his enemies invite him to arbitrate on questions of good or evil, he rejects 'the human either-or that is implied in every such question' (p. 312).[83] His life is lived in the true freedom of simplicity and single-mindedness: 'He lives and acts not out of knowledge of good and evil, but out of the will of God. There is only *one* will of God. In it, the origin has been regained' (p. 313, emphasis original). Likewise, for his followers, 'those for whom the knowledge of good and evil has been overcome, there is no longer a choice among various possibilities but always only the one option of being elected to do the one will of God in simplicity' (p. 320). So, against the preoccupation with good and evil that makes us 'essentially judges', Jesus says 'Do not judge, so that you may not be judged' (Mt. 7.1; pp. 313–14). Against our self-conscious knowledge of the good and evil we have done, he says 'do not let your left hand know what your right hand is doing' (Mt. 6.3) – so that in the parable of the sheep and the goats (Mt. 25.31-46), the righteous who have fed, clothed and cared for Jesus 'will not know their own goodness; Jesus will reveal it to them' (p. 319).

We should not imagine, however, that our '[election] to do the one will of God in simplicity' puts an end to thought and reasoning. For those in whom the knowledge of good and evil has been overcome by Christ, there is the task of discernment signalled by Paul's injunction to 'be transformed by the renewing of your minds, so that you may discern what is the will of God – what is good and acceptable and perfect' (Rom. 12.2). For those set free from being 'essentially judges', there is instead 'a judging that is a genuine human activity, that is, a "judging" that springs from the accomplished unity with the origin, with Jesus Christ' (p. 316). In Paul's words, 'Those who are spiritual judge all things, and they are themselves judged by no one' (1 Cor. 2.15, translation adapted from NRSV).[84] Alongside Jesus's saying 'do not let your left hand know what your right hand is doing', there is Paul's admonition, 'Examine yourselves to see whether you are living in the faith. Test yourselves. Do you not realize that Jesus Christ is in you?' (2 Cor. 13.5). In contrast to the kind of self-examination that 'focus[es] on one's

[83] Bonhoeffer quotes Lk. 12.14: 'Friend, who set me to be a judge or arbitrator over you?'

[84] NRSV reads 'Those who are spiritual discern all things, and they are themselves subject to no one else's scrutiny.' Both Greek verbs in this sentence are forms of *anakrinō*.

own knowledge of good and evil and its realization in practical life', says Bonhoeffer, what Paul commends is a Christian self-examination that 'daily renews the knowledge that "Jesus Christ is in us"' (p. 325). This discernment of God's will is a concrete task that must result in the actual *doing* of God's will. It is also a daily task, to be undertaken anew in each new situation in which we find ourselves.

This may seem like the worst kind of voluntarism and situationism, but other fragments of the *Ethics* make it clear that God's will is not to be understood in an arbitrary or situationist manner. Nor is its discernment a task that bypasses human reason and reflection:

> Intellect, cognitive ability, and attentive perception of the context come into lively play here. All of this discerning will be encompassed and pervaded by the commandment. Prior experiences will raise encouraging or cautionary notes. Under no circumstances must one count on or wait for unmediated inspirations, lest all too easily one fall prey to self-deception … Possibilities and consequences will be considered carefully. In short, in order to discern what the will of God may be, the entire array of human abilities will be employed (pp. 323–24).

Here Bonhoeffer shows himself alive to the danger of being misled by intuitive judgements (which, for believers, may take on the appearance of insights that come by direct divine inspiration) and – in *one* sense rather like Greene – recommends the careful use of reason to avoid this danger. The crucial difference from Greene is that for Bonhoeffer, what is needed is theologically informed reason, 'encompassed and pervaded by the [divine] commandment' and directed to the task of 'discern[ing] what the will of God may be'. So do the scientific insights discussed in the present chapter have any part to play in the task of discerning God's will, so understood?

It might appear not, because Bonhoeffer draws a sharp distinction between psychological and theological perspectives. His account of freedom and simplicity, he says, would 'be completely misunderstood if these things were taken to be psychologically observable facts' (p. 319). Because 'psychological observation itself is always already subject to the law of disunion … [it will] never be able to discover the simplicity, the freedom, and the doing that Jesus intended' (p. 320). We will not, then, learn from the psychological or cognitive sciences that the ethical task is to do the one will of God in simplicity and freedom, nor will those sciences tell us what that simplicity and freedom look like.

Bonhoeffer's tone appears rather dismissive of the psychological sciences – perhaps more than he should be, even by his own lights. He does, after all, go on to say that 'the entire array of human abilities' should be employed in the task of discernment (p. 324). Similarly, Barth, every bit as insistent as Bonhoeffer that scientific knowledge in itself cannot discover the reality of the human creature, nonetheless allows that within its own limits it can offer genuine insights.[85] Scientific and philosophical accounts of the 'phenomena of the human' are, he remarks, 'like an interesting commentary on a text which must first be known and read for itself if the commentary is to be intelligible and useful'.[86] A commentary can help us understand a text better, provided the commentator has taken the trouble to be faithful to the text. So if we know on theological grounds that the ethical task is to do the one will of God in simplicity and freedom, neuroethical commentary on the human 'text' could in principle help us understand better what it means for concrete human lives to fulfil that task. To be sure, putative insights from neuroscience are at best partial, fallible and provisional, and they may be distorted in the various ways discussed earlier in the chapter. So they must be *critically* appropriated into a theological ethic, with all due caution and keen awareness of their limitations. Nonetheless, the appropriate response is critical reception rather than simple rejection.

To understand how this might work in practice, we may revisit Joshua Greene's fullest treatment of neuroethics to date, which concludes with 'six rules for modern herders' – that is, for us who must coexist not only with members of our own 'moral tribe' but with other 'tribes' whose commonsense moralities differ from ours.[87] First, in our personal lives, intuitive moral judgements are more reliable guides than reasoning, which is susceptible to unconscious self-interested bias; but when faced with controversies in which the intuitive judgements of different moral tribes conflict, we must resort to reason. Second, rights language (the main deontological approach

[85] Barth, *Church Dogmatics*, vol. 3.2, pp. 71–132.
[86] Ibid., p. 122.
[87] Greene, *Moral Tribes*, pp. 350–52.

he considers) should be used sparingly and recognized for what it is: not a rationally defensible moral theory but a rhetorical weapon. Third, in moral controversy, we should focus on the facts: on empirical evidence about what will increase benefit and reduce harm. Fourth, we should beware of 'biased fairness': the unconscious tendency to select whichever version of 'fairness' suits our own interests. Fifth, to settle moral controversies, we should use the 'common currency' of utilitarian reasoning. And finally, we (the implied audience is more or less affluent Westerners) should give to charity, because this can make a dramatic difference to the well-being of others and partially overcome our intuitive indifference to strangers.

The Bonhoeffer-informed theological account that I have proposed might respond to Greene's six practical rules along the following lines. Given its critique of human pretensions to the knowledge of good and evil, this theological perspective will share Greene's suspicion of rights language and other deontological approaches insofar as these are seen as attempts to provide a universal, rationally defensible basis for moral norms.[88] This is not to say that such moral language should be rejected wholesale. In another fragment of *Ethics*, for example, Bonhoeffer develops an extensive account of natural rights, but it is his *theological* understanding of the natural that enables his rights-talk to be much more than mere rhetorical weaponry.[89] This Bonhoefferian perspective will also share Greene's concerns about biased fairness – our tendency to spin our moral principles to our own advantage – and the tendency for our moral reasoning to be deployed in self-interested and self-deceiving ways.

Indeed, in some respects, this perspective will both intensify and extend the range of Greene's suspicion. In particular, theological suspicion of ethics must also be directed at Greene's *own* moral reasoning and the status of 'metamorality' that he claims for utilitarianism. He fails to acknowledge that utilitarianism also depends on a historically and culturally particular set of presuppositions

[88] However, some of Greene's philosophical critics point out that not all deontological ethics answer to this description: see, for example, Mark Timmons, 'Toward a Sentimentalist Deontology', in Sinnott-Armstrong (ed.), *Moral Psychology, Vol. 3*, pp. 93–104.

[89] Bonhoeffer, *Ethics*, pp. 171–218. For a recent theologically-grounded account of rights language, and a critique of rights discourse that aspires to tradition-neutral universality, see Esther D. Reed, *The Ethics of Human Rights: Contested Doctrinal and Moral Issues* (Waco, TX: Baylor University Press, 2007); idem, 'Responsibility to Protect and Militarized Intervention: When and Why the Churches Failed to Discern Moral Hazard', *Journal of Religious Ethics* 40.2 (2012), pp. 308–34 (325–29).

about the human good.[90] He could have learned this from Nietzsche (given his readiness to use him as an ally in other contexts), since the heart of Nietzsche's critique of utilitarianism was that it was a secularized version of the Christian slave-morality he so despised.[91] Greene's theological critics may be less worried about any lingering influence of Christianity on utilitarianism, but will have concerns about what has been lost in the process of secularization, particularly in the reductive redefinition of the human good as pleasure and the absence of pain (or other formulations of utility). The point is that such conceptions of the good are shaped by the particular historical, cultural and social milieux from which they emerged, as critics such as Charles Taylor demonstrate,[92] and have no more convincing claims to universal validity than the deontological views Greene debunks. Of course Greene emphasizes that he does not claim utilitarianism discloses an objective, universally valid moral truth, merely that it is a pragmatic way to find a 'common currency' (harm and benefit) for assessing the claims of rival communities with incompatible moralities.[93] But even this more modest kind of universal claim is suspect, because if 'utility' reflects a *particular* moral vision and understanding about what is of value in human life, we cannot be confident that the unit of value it supplies will be agreed upon by those who hold different moral visions. In short, Greene's attempt to promote utilitarianism as a metamorality looks more like a land-grab by one moral tribe than a tribe-neutral way of enabling everyone to coexist on the new pastures.

It follows that Greene's third rule about focusing on the facts also requires severe qualification. It has an appearance of objectivity, of taking moral controversies out of the realm of unprovable theory into the domain of hard empirical evidence. In reality, though, Greene's unacknowledged and contestable presuppositions about the human good generate dubious claims about what the relevant facts *are*. This is evident in his extended example of the abortion debate, which includes various questionable moves.[94] First, he tries to 'call the bluff' of rights-based arguments on both sides, arguing that they depend on intuitive judgements that are hard to defend rationally

[90] See Charles Taylor, *Sources of the Self: The Making of the Modern Identity* (Cambridge: Cambridge University Press, 1989), ch. 19.

[91] Anomaly, 'Nietzsche's Critique of Utilitarianism'.

[92] Taylor, *Sources of the Self*, ch. 19.

[93] Greene, *Moral Tribes*, pp. 188–89.

[94] Ibid., pp. 309–27.

and consistently. In doing so he makes some contestable claims, for example, that pro-life arguments depend on unverifiable assertions about ensoulment (some do, perhaps, but *all*?). Having rejected rights-based arguments he turns to utilitarianism, apparently presuming that no other non-utilitarian arguments need be considered. He dismisses a pro-life argument based on the happiness that would have resulted from the lives of aborted individuals had they lived, because it entails an absurd obligation to bring as many new lives as possible into being. We cannot factor in the happiness of hypothetical people who will never live, he argues, so what remains on the balance sheet is the happiness that will be lost by '[d]isrupting people's sex lives, disrupting people's life plans, and forcing people to seek international or illegal abortions'.[95] However, in dismissing the claims of hypothetical human individuals who will never live, he has also dismissed the claims of human individuals that already exist, but *in utero*. The individuals whose happiness appears on the balance sheet are those at a stage of development when they can have sex lives, life plans, etc. In other words, some of the work in Greene's argument is being done by a covert assumption about what kinds of human individual have interests that merit consideration, and what kinds do not. But of course that is precisely what is at issue between pro-life and pro-choice camps; a central question has been begged. My main point here is not to advocate a position in the abortion debate as such, but to suggest that if utilitarianism is supposed to be an impartial common currency for handling such intractable conflicts, on the strength of this example it does not inspire much confidence.

The theological account I have proposed will not, of course, dismiss empirical evidence: according to Bonhoeffer, discerning God's will requires attention to past experience, 'attentive perception of the context', and consideration of the likely consequences of one's decisions, among other things.[96] We shall, however, need to be very clear about the limits of what empirical evidence can tell us. On some questions it might form a significant part of the argument, but on others its role would be more limited. Debates about legalizing assisted dying, for example, will hardly be settled by evidence of the effects of legislation in Oregon or the Netherlands, although such evidence is not irrelevant.

95 Ibid., p. 326.

96 Bonhoeffer, *Ethics*, pp. 323–24.

I have contested Greene's claims for utilitarianism as a metamorality. More generally, a theological suspicion of the knowledge of good and evil should make us sceptical about *any* quest for a metamorality, or overview that would enable us to settle the controversies between contrasting moral visions. This sceptical point can be explained as follows. As I suggested earlier, Greene's framing of the problem of 'Us versus Them' echoes Alasdair MacIntyre's observation about the fragmentation of modern moral discourse.[97] MacIntyre and Greene describe the symptoms in similar ways, but offer different diagnoses and remedies. For MacIntyre the incoherence is a product of modernity, it is exposed by Nietzsche and the solution is a recovery of the more coherent Aristotelian (and in later works, Thomist) moral tradition that flourished before modernity. For Greene, the incoherence becomes *evident* in modernity, and like MacIntyre, he enlists Nietzsche as a critic to help him demonstrate it (though unlike MacIntyre, he tries to deploy Nietzsche selectively, against only some forms of modern ethics).[98] But for Greene the roots of the problem are not in the Enlightenment project, but lie much deeper, in the evolutionary origins of morality itself. For him, therefore, the premodern tradition of the virtues is not the solution, but an aspect of the problem.[99] Yet he is more sanguine than MacIntyre about the capacity of modern ethics to solve the problem. The theological analysis I have drawn from Bonhoeffer will be more radical than either Greene or MacIntyre.[100] It will agree with MacIntyre that any attempt to overcome the incoherence of moral discourse using modern ethical theory is futile, but it will concur with Greene that the problem is as old as humanity, so the solution will not be found in premodern philosophical traditions either. For Bonhoeffer, the root of the problem is humanity's 'falling away from the origin' into the disunion of the knowledge of good and evil.[101] Any attempt to overcome that disunion by reasoning about the knowledge of good and evil (in other words, to devise what Greene calls a metamorality) is doomed to fail. What we need is to 'regain' the origin through Jesus Christ.[102]

97 MacIntyre, *After Virtue*, ch. 1; Greene, *Moral Tribes*, pp. 382–83.
98 Greene, 'Secret Joke', pp. 36–79; *Moral Tribes*, pp. 301, 331, 332.
99 He describes Aristotelian virtue ethics as a tribal philosophy, oddly citing MacIntyre in support: *Moral Tribes*, p. 330.
100 'Radical', that is, in the general sense of getting to the root of the matter, not in Bonhoeffer's specialized and critical sense denoting the erroneous neglect of the 'penultimate' for the sake of the 'ultimate': *Ethics*, pp. 153–57.
101 Bonhoeffer, *Ethics*, p. 300.
102 'In [the one will of God], the origin has been regained': ibid., p. 313.

As I noted earlier, Greene could no doubt respond that this analysis reflects one particularly idiosyncratic tribal morality, and is therefore part of the problem with little to contribute to the solution. I acknowledged earlier that Greene's scientifically-based critique may be a salutary reminder that the Church and its ethics too stand under judgement. However, this theological welcome for Greene's critical perspective has its limits. Those who take Bonhoeffer's view are in effect framing the problem differently from Greene, and need not agree with an objection premised on his way of framing it – particularly when the debate moves on to the merits of his proposed solution. Theologians who take this view are, though, bound to say something about how a Christian ethic which is 'the critique of all ethics' should engage with an estranged and disunited world.

One way to do so is of course by witnessing to the Gospel: the good news that 'regaining the origin' is possible in Christ. But there is more to be said. If this good news is the 'ultimate', the last word that God has spoken to humanity in and through Christ, then it places on Christians a weighty responsibility for the *pen*ultimate: the here-and-now world, the place where God comes to us with a word of salvation in Christ.[103] It is this responsibility for the penultimate in the light of the ultimate that grounds Bonhoeffer's account of natural life and human rights.[104] And this responsibility for the penultimate mandates Christians to find ways of collaborating with others to care for the world and human life. So we must, among other things, find ways of working with differences of moral vision and conviction. But if Bonhoeffer is right about the knowledge of good and evil, trying to deal with such differences by seeking a 'metamorality' will be liable to distort and mislead, as I have already suggested. A more promising model for working with these differences is *conversation*.[105] This does not require an allegedly neutral common moral currency for weighing the values and convictions of different moral tribes, but merely a shared commitment to speak and listen to one another and make a serious effort to understand each other's perspectives. It need not have grandiose aims, but can have the more modest goal of reaching limited agreement on particular

[103] Bonhoeffer, 'Ultimate and Penultimate Things', in *Ethics*, pp. 146–70.

[104] Bonhoeffer, 'Natural Life', in *Ethics*, pp. 171–218.

[105] Cf. Nigel Biggar, *Behaving in Public: How to Do Christian Ethics* (Grand Rapids, MI: Eerdmans, 2011), pp. 63–78.

questions. An openness to conversation of this sort should be accompanied by a commitment to collaborative *action* with a wide range of others in pursuit of shared goods, collaboration which again might be specific, ad hoc and limited in its aims and goals.[106]

Lastly, this theological account will of course endorse Greene's final precept, to give charitably as a way of partially overcoming the inbuilt weakness of our sympathy with distant strangers. It will, however, press his point further, calling for more far-reaching practices of hospitality to strangers and solidarity with the poor and marginalized, wherever in the world those neighbours might be.

Conclusion

Among its many conversation partners, there is every reason for theological ethics to attend carefully and critically to the kinds of scientific study of morality discussed in this chapter. This attention must be critical, because the difficulties noted earlier with the conceptualization, design and interpretation of studies such as Greene's mean that even the scientific conclusions he and his colleagues draw are certainly not beyond question. But attention that is also careful and receptive could pay dividends, because Greene and his colleagues – like Feuerbach and the other 'masters of suspicion' before them – may prove friends in disguise to theological ethics. Greene's critical perspective on what he calls our 'point-and-shoot' moralities and the theoretical rationalizations we construct for them is a timely reminder to theologians to have a properly *theologically* critical view of the human project of ethics. On this, as on other topics, neuroscience could offer helpful commentary that (used judiciously) may be of real value to theological readers of the 'text' of human creaturely being.

[106] Cf. Luke Bretherton, *Christianity and Contemporary Politics* (Chichester: Wiley-Blackwell, 2010), esp. ch. 2. There may be some tension between the commitments to conversation and collaborative action (as there is between Biggar's and Bretherton's perspectives on them), but for the most part that tension should prove creative rather than destructive. See further Kathryn Pritchard, 'Bioethics, Public Policy and the Church of England' (unpublished doctoral dissertation, University of Winchester, 2015).

4

Freedom, Responsibility, Sin and Grace: 'Mr Puppet' Meets St Augustine

Introduction

Are all our choices and actions caused by physical processes in the brain? If so, does that mean they are not truly free? Moreover, are there psychological grounds for thinking that our experience of having free will is an illusion? If we do not have free will, does that also mean we cannot be responsible for our actions? In short, do we deceive ourselves if we imagine that we are free and responsible agents?

Questions like these have long been debated in one form or another, but in recent decades they have been given a new twist by research in neuroscience and psychology. First, studies by Benjamin Libet and his colleagues in the 1980s suggested that voluntary actions are initiated by unconscious brain activity before we become aware of consciously willing them. If this is so, it seems that our conscious willing cannot be the cause of our actions. More recent neuroscientific studies have built on Libet's work, developed it and improved on it in various ways. Alongside this, the psychologist Daniel Wegner has argued that our experience of conscious will is an illusion, an 'authorship emotion' by which the brain labels certain actions as ours and associated them with appropriate moral emotions such as pride or shame. Other areas of psychological research, while not calling our free will and agency into question wholesale in this way, have suggested that it may be more limited than we imagine. These scientific studies have been extensively argued over by philosophers of mind, some arguing that they radically undermine our view of ourselves as free and responsible agents, others maintaining that they pose little threat to that self-understanding.

In the first half of this chapter, I shall survey some of these neuroscientific and psychological accounts and a selection of philosophical arguments about them. In particular, I shall discuss four areas of debate: determinism and free will, questions about mental causation, some particularly perplexing questions about addiction and autonomy, and finally the implications of all this for our notions of moral responsibility.

This topic has not only been extensively argued about by scientists and philosophers: it is also something of an exception to my complaint in Chapter 1 about theologians' neglect of neuroethics. Perhaps because debates about neuroscience, freedom and responsibility long pre-date the emergence of neuroethics as a field, they have received more theological attention than many neuroethical topics. In the second half of this chapter, I shall draw on some of these theological resources when I turn to a theological engagement with the questions surveyed in the first half. However, I shall go further than some previous theological discussions, and attempt a reframing of those questions by relocating them in the context of core Christian doctrines: specifically, divine providence, sin and salvation. As we shall see, some basic assumptions found in the scientific and philosophical debates may require revision if we try to think theologically about human agency, freedom and responsibility in this way. However, I shall argue that such rethinking has the potential to shed new light on sometimes perplexing and troubling questions about what it means to be a free and responsible human being.

Definitions and distinctions

Neuroscientific and philosophical discussions of free will can become rather dense and technical. It may help readers navigate these discussions if I first offer some definitions of key terms and concepts. Note, however, that these are only initial definitions; in some cases, they incorporate assumptions that may require challenging and rethinking as our exploration proceeds.

To begin with *free will*: according to one neuroethics text, it is traditionally 'conceptualized as the capacity possessed by persons to decide and act in accordance with an unimpeded will of their own'.[1] Echoing a widely (though

[1] Bernard Baertschi and Alexandre Mauron, 'Genetic Determinism, Neuronal Determinism, and Determinism *Tout Court*', in Judy Illes and Barbara J. Sahakian (eds.), *The Oxford Handbook of Neuroethics* (Oxford: Oxford University Press, 2011), pp. 151–60 (155).

not universally) held assumption, the authors continue: 'The criterion of an unimpeded will is the possibility for it to choose between different alternatives.'[2] But what is required for a will to count as unimpeded or free? Different concepts of freedom set the bar at different levels. In his *Treatise of Human Nature*, David Hume made a famous distinction between *liberty of spontaneity* ('that which is oppos'd to violence') and *liberty of indifference* ('that which means a negation of necessity and causes').[3] If by freedom we mean the liberty of spontaneity, a choice or action is free if it is not compelled, coerced or constrained. But if we mean the liberty of indifference, then to count as free a choice or action must meet the more stringent requirement of not being (completely) causally determined.

This brings us to *causal determinism* and its relationship to free will. A standard definition of determinism is that 'a complete description of the state of the world at a particular time, in conjunction with a complete formulation of the natural laws, entails the truth about the physical state of the world at later times'.[4] Is causal determinism, so defined, compatible with free will? In the standard philosophical terminology, *compatibilists* think it is, *incompatibilists* deny this. Incompatibilists are subdivided into *hard determinists*, who hold that determinism is true and reject free will, and *libertarians*, who affirm free will and reject determinism.[5] Notice that these positions tend to be associated with different conceptions of freedom. For many compatibilists, choices and actions are genuinely free if something like the liberty of spontaneity applies to them. Incompatibilists are likely to insist that a choice or action is only truly free if the liberty of

2 Ibid.; so also Joshua Greene and Jonathan Cohen, 'For the Law, Neuroscience Changes Nothing and Everything', in Martha J. Farah (ed.), *Neuroethics: An Introduction with Readings* (Cambridge, MA: MIT Press, 2010), pp. 232–58 (236). However, some compatibilist philosophers challenge the assumption that free will requires alternative possibilities: see, for example, Harry Frankfurt, 'Alternate Possibilities and Moral Responsibility', *Journal of Philosophy* 66 (1966), pp. 829–39; Walter Glannon, *Brain, Body and Mind: Neuroethics with a Human Face* (Oxford: Oxford University Press, 2011), p. 44.

3 David Hume, *A Treatise of Human Nature* (L. A. Selby-Bigge and P. H. Nidditch (eds.); Oxford: Clarendon, 1978), p. 407.

4 Glannon, *Brain, Body and Mind*, p. 44.

5 See, for example, Daniel C. Dennett, *Freedom Evolves* (London: Allen Lane, 2003), pp. 97–98. A further possible position is that whether or not determinism is true, there cannot be free will: Galen Strawson, 'The Bounds of Freedom', in Robert Kane (ed.), *The Oxford Handbook of Free Will* (Oxford: Oxford University Press, 2002), pp. 441–60.

indifference – at least an element of genuine causal indeterminacy – can be attributed to it.

Distinct from causal determinism is the question of *mental causation*: can mental events be causes of physical events, and in particular, can conscious volitions cause physical actions? Philosophical debates about free will have focused largely on determinism, but the neuroscientific work discussed in this chapter also raises questions about mental causation, independently of what one says about determinism and free will.[6]

Nancey Murphy has drawn attention to the role of *reductionism* in the problem of mental causation.[7] Reductionism in general involves understanding the world in terms of different levels of complexity, and regarding the lowest or simplest level as 'metaphysically basic'.[8] There are different varieties:[9] for example, *methodological* reductionism is a research strategy of investigating how a system works by analysing it into its component parts. *Ontological* reductionism sometimes refers to the thesis that in considering increasingly complex entities, 'no new ontological "ingredients" need to be added'[10] – there is no need to postulate a 'vital force' in order to describe living systems, for example. It may also refer to the stronger claim that 'only the entities at the lowest level of the hierarchy are "*really* real"'.[11] The variety most relevant to the problem of mental causation is *causal reductionism*: the thesis 'that behavior of parts of a system … is determinative of the behavior of all higher-level entities', or 'that all causation in the hierarchy is "bottom-up"'.[12] As Murphy observes, this is closely related to the stronger version of ontological reductionism.

Having defined some of the key terms and concepts, at least in a preliminary way, we are ready to begin exploring the science and the questions it raises.

6 Cf. Walter Sinnott-Armstrong and Lynn Nadel, 'Introduction', in Walter Sinnott-Armstrong and Lynn Nadel (eds.), *Conscious Will and Responsibility: A Tribute to Benjamin Libet* (Oxford: Oxford University Press, 2010), pp. xi–xv.

7 Nancey Murphy, 'Avoiding Neurobiological Reductionism: The Role of Downward Causation in Complex Systems', in Juan José Sanguineti, Aribierto Acerbi and José Angel Lombo (eds.), *Moral Behavior and Free Will: A Neurobiological and Philosophical Approach* (Morolo: If Press, 2011), pp. 201–22.

8 Carlos J. Moya, 'Mind, Brain, and Downward Causation', in Sanguineti et al. (eds.), *Moral Behavior and Free Will*, pp. 185–200 (188).

9 Murphy, 'Avoiding Neurobiological Reductionism', pp. 202–3.

10 Ibid., p. 203.

11 Ibid.

12 Ibid.

Neuroscientific questions and philosophical debates

This section of the chapter begins with Libet's neuroscientific studies about conscious will and action. In the first subsection, I survey some of the criticisms of Libet's work, and some of the more recent work by researchers who have attempted to build and improve upon Libet. I explore various arguments about what this work might, or might not imply, for two questions that I identified in the introduction: the question of mental causation and the issue of determinism and free will. In the next subsection, the focus shifts to areas of psychological research that do not challenge free human agency wholesale, but may constitute what Thomas Nadelhoffer calls 'partial agential threats' by showing that the scope of consciously willed action is much more limited than we often suppose. After that, I turn to some recent neuroethical discussion about addiction and the questions it raises about autonomous freedom. In the final subsection, I ask what might be the implications of all this for our notions of moral responsibility.

Is free will an illusion?

What does a scientific understanding of causation imply for freedom and responsibility? The discussion of this long-debated question took a new turn in the 1980s thanks to the work of Benjamin Libet and his colleagues.[13] Their research built on an earlier finding that voluntary actions were preceded by a characteristic change known as the 'readiness potential', detectable by electroencephalography (EEG), which could begin up to a second before activation of the relevant muscle. In Libet's study, subjects were asked to perform a simple action (a flick of the wrist) at whatever time they felt the urge to do so. The readiness potential and the time of muscle activation were measured, and subjects were asked to report the time when they first became aware of the wish to flick their wrist. Libet et al. found that the readiness potential began on average 550 milliseconds before the muscle was activated. The time when subjects were first aware of wishing to perform the action was

[13] An account and discussion of this work can be found in Benjamin Libet, 'Do We Have Free Will?' *Journal of Consciousness Studies* 6 (1999), pp. 47–57. For an extensive and valuable collection of recent essays on the issues raised by Libet's work, see Sinnott-Armstrong and Nadel, *Conscious Will and Responsibility*. Libet's essay is reprinted as ch. 1 of this collection.

on average 150 milliseconds before muscle activation began, or around 400 milliseconds *after* the beginning of the readiness potential.

Libet concluded that this result cast doubt on commonsense notions of free will in which conscious willing is the cause of the willed action. If the action is initiated by unconscious brain activity before the time of the conscious volition (as his study appeared to show), it would seem that the conscious volition cannot be the cause of the action. Libet argued that conscious free will could still play a role, because the 150 millisecond gap between awareness of the wish to act and the activation of the muscle allowed time for a conscious *veto* of the unconsciously initiated action. To repeat a common wisecrack, we may not have free will, but we still have 'free won't'.

Critics have raised various questions about the design and conceptualization of Libet's experiments.[14] First, there are technical concerns. For example, since the readiness potential was measured at the scalp, it is not clear exactly where in the brain the signal was generated. Also, the causal relationship between readiness potentials and actions is not as clear as Libet assumed, and there is some evidence that readiness potentials are neither necessary nor sufficient for voluntary muscle movements.[15] A third set of technical concerns relates to the subjects' reports of the time at which they became conscious of the urge to act. These reports may be unreliable and highly sensitive to details of the experimental design, such as the type of clock used.[16]

Questions have also been raised about the conceptualization of Libet's studies: was he measuring what he thought he was? For example, his subjects were asked to report on the 'urge' or 'wish' to flick their wrists, but an urge is not the same thing as a decision. There is some evidence that decisions may not be associated with the same pattern of readiness potentials as urges.[17] More broadly, Roskies suggests that when subjects reported awareness of the urge to act, they were not reporting a conscious volition itself, but a 'meta-state',

[14] For the following points see, for example: Adina L. Roskies, 'Why Libet's Studies Don't Pose a Threat to Free Will', in Sinnott-Armstrong and Nadel (eds.), *Conscious Will and Responsibility*, pp. 11–22; Susan Pockett and Suzanne C. Purdy, 'Are Voluntary Movements Initiated Preconsciously? The Relationships between Readiness Potentials, Urges, and Decisions', in Sinnott-Armstrong and Nadel (eds.), *Conscious Will and Responsibility*, pp. 34–46; Glannon, *Brain, Body and Mind*, ch. 2.

[15] Pockett and Purdy, 'Are Voluntary Movements Initiated Preconsciously?' pp. 35–37.

[16] William P. Banks and Eve A. Isham, 'Do We Really Know What We Are Doing? Implications of Reported Time of Decision for Theories of Volition', in Sinnott-Armstrong and Nadel (eds.), *Conscious Will and Responsibility*, pp. 47–60.

[17] Pockett and Purdy, 'Are Voluntary Movements Initiated Preconsciously?', pp. 39–44.

consciousness of the conscious volition. In that case, its timing would tell us little about the timing of the conscious volition itself, and it might be quite possible to have consciously willed acts without this 'meta-intentional' state.[18] She also argues that what really matters about free will is that we are free to act for salient reasons. In that perspective, '[a]rbitrary action is, at best, a degenerate case of freedom of the will, one in which what matters about freedom fails to hold'.[19] A related point is made by several critics, including Walter Glannon: by focusing on a very narrow slice of time, Libet's experiments (and indeed more recent neuroscientific studies) may capture only part of the mind and brain activity leading to the action. Libet may have captured subjects' *proximal* intentions to flick their wrists, but not their longer-lasting (and conscious) *distal* intentions to participate in his experiments and obey his instructions.[20] Roskies' and Glannon's critiques suggest that what Libet's experiments tell us about willed action in the real world may be limited at best – which he himself acknowledged up to a point.[21]

Some of these critical points have been addressed by more recent studies. For example, Mark Hallett has described a variant of the Libet experiment which does not rely on subjects' reports of the timing of their wishes to act or their actions. Like Libet, he found that unconscious brain activity occurred earlier than conscious volitions, and by contrast with Libet, Hallett's results did not suggest there was sufficient time for a conscious veto of unconsciously-initiated actions.[22] In fMRI studies by John-Dylan Haynes and colleagues, specific patterns of activity predicted more accurately than chance which of two actions a subject would choose, around 7–10 seconds before the action took place.[23] (Hallett's and Haynes' studies, like Libet's, focused on simple arbitrary actions with little salience for their subjects, so Roskies' query about relevance to real world decision-making applies to them too.)

[18] Roskies, 'Why Libet's Studies Don't Pose a Threat', pp. 20–22.

[19] Ibid., p. 18.

[20] Glannon, *Brain, Body and Mind*, p. 53.

[21] Libet, 'Do We Have Free Will?' pp. 53–54.

[22] Masao Matsuhashi and Mark Hallett, 'The Timing of the Conscious Intention to Move', *European Journal of Neuroscience* 28 (2008), pp. 2344–51; Mark Hallett, 'Volition: How Physiology Speaks to the Issue of Responsibility', in Sinnott-Armstrong and Nadel (eds.), *Conscious Will and Responsibility*, pp. 61–69.

[23] Specifically, in the frontopolar cortex and the precuneus/posterior parietal cortex: John-Dylan Haynes, 'Beyond Libet: Long-Term Prediction of Free Choices from Neuroimaging Signals', in Sinnott-Armstrong and Nadel (eds.), *Conscious Will and Responsibility*, pp. 85–96.

As I suggested in the chapter introduction, these studies raise two distinct questions: about determinism (whether an action can be free if it is the product of physical causes) and mental causation (whether mental events such as conscious volitions can and do cause physical events).

To begin with the second, Daniel Wegner maintains that Libet's results, together with a range of other evidence, demonstrate that conscious will is an illusion.[24] His other evidence includes situations in which the experience of conscious will is separated from action: positive illusions of control, on the one hand, and actions that we do not experience as willed, on the other. In his theory of 'apparent mental causation', volition is an 'authorship emotion' that tags certain events as ours,[25] leading us to *interpret* thoughts as causing actions. The mind is most likely to do this when the thoughts occur shortly before the actions, are consistent with them, and when it is not aware of likely looking alternative causes (internal or external).[26] In experiments that replicate these conditions, subjects erroneously experience their conscious will as having caused events. In Wegner's theory, 'conscious will, and authorship more generally, is less a cause of events than an *experience* one has when the mind determines an event should be ascribed to the self'.[27] The mind maintains the illusion of conscious will by various mechanisms: for example, confabulating intentions and distorting perceptions of the time between actions and events.[28]

One challenge to Wegner comes from Terry Horgan,[29] who argues that the experience Wegner thinks is illusory is of a *mental state* (such as the wish to move one's wrist) causing a bodily movement. Horgan maintains this is different from the phenomenology of willed action, in which we experience *ourselves* as the cause. It is more economical, he thinks, to explain that experience (of ourselves as the cause) as veridical than as a confabulation. His disagreement with Wegner will turn, at least in part, on the reliability of the particular kinds of introspection to which he appeals. Horgan believes handled carefully,

[24] Daniel M. Wegner, *The Illusion of Conscious Will* (Cambridge, MA: MIT Press, 2002).

[25] Ibid., ch. 9.

[26] Ibid., ch. 3.

[27] Jeffrey P. Ebert and Daniel M. Wegner, 'Bending Time to One's Will', in Sinnott-Armstrong and Nadel (eds.), *Conscious Will and Responsibility*, pp. 134–45 (135, emphasis original).

[28] Wegner, *The Illusion of Conscious Will*, ch. 5; Ebert and Wegner, 'Bending Time to One's Will'.

[29] Terry Horgan, 'The Phenomenology of Agency and the Libet Results', in Sinnott-Armstrong and Nadel (eds.), *Conscious Will and Responsibility*, pp. 159–72.

introspection can yield reliable information on *some* aspects of agency,[30] which Wegner would seem much more inclined to doubt.

While Wegner thinks his theory has only modest implications for our view of agency, others see it as more far-reaching. According to Thomas Nadelhoffer, it poses a 'global' threat to our self-understanding as moral agents, seeming to 'leave us adrift on a sea of unconscious forces that we are powerless to change'.[31] He believes this threat is distinct from the one posed by determinism, and arises irrespective of one's position on the latter.

The experiments of Libet, Haynes and others focus on experimental situations with debatable relevance for decision-making and acting in the real world. Some authors, however, go further, expressing global scepticism about the possibility of mental causation. For example, Jaegwon Kim defends a causally reductionist view in which mental states supervene on physical brain states but have no causal relationship to other physical or mental states.[32] The alternatives, he thinks, are that both physical and mental states could be sufficient causes of a new physical brain state, or that a mental state could exercise downward causation on the physical state of the brain. He rejects the first of these because it would mean that the physical state of the brain is causally overdetermined, and the second because it would violate the causal closure of the physical world.

Against Kim, Nancey Murphy proposes an account of top-down causation drawing on the work of Terrence Deacon.[33] In this account, complex physical systems demonstrate new *emergent properties*: physical properties that result from interactions between the constituent parts of the system, but only appear once a certain level of complexity is reached. Different levels of emergence can be identified. At the most complex, a variety of self-organizing forms emerge, and their interactions with the environment determine which of them will

[30] Ibid., p. 161.

[31] Thomas Nadelhoffer, 'The Threat of Shrinking Agency and Free Will Disillusionism', in Sinnott-Armstrong and Nadel (eds.), *Conscious Will and Responsibility*, pp. 173–88 (183).

[32] For example, Jaegwon Kim, 'Blocking Causal Drainage and Other Maintenance Chores with Mental Causation', *Philosophy and Phenomenological Research* 67.1 (2003), pp. 151–76; see also Kim, *Mind in a Physical World* (Cambridge, MA: MIT Press, 1998).

[33] Murphy, 'Avoiding Neurobiological Reductionism', citing Terrence W. Deacon, 'Three Levels of Emergent Phenomena', in Nancey Murphy and William R. Stoeger (eds.), *Evolution and Emergence: Systems, Organisms, Persons* (Oxford: Oxford University Press, 2007), pp. 88–110. See also Nancey Murphy and Warren S. Brown, *Did My Neurons Make Me Do It? Philosophical and Neurobiological Perspectives on Moral Responsibility and Free Will* (Oxford: Oxford University Press, 2007).

persist into the future. One example is evolution by natural selection. Another might be the process of 'synaptic refinement' in childhood and adolescence. The number of synaptic connections in human brains peaks between nine and eleven years of age. As the brain continues to mature, the number decreases, as some are lost and others maintained because of inputs from the body and the environment.[34] According to Murphy, this is one example of how mental processes can exert downward causation on physical states or processes in the brain. In this account, top-down causation operates, not by altering the lower-level causal processes (such as cell growth and the formation of synaptic connections), but by constraining and selecting which of them take effect. Walter Glannon suggests that another example is cognitive behavioural therapy, where reframing beliefs and habits of thought can modulate physical states and functions of the brain.[35]

So far I have been discussing the question of mental causation. The other question raised by Libet's results was about determinism: if conscious volitions are caused by physical processes in the brain, can they be truly free? This question, of course, has long been debated by philosophers of mind; but in the hands of some authors, old debates about determinism have been given a new twist by neuroscience.

For example, Joshua Greene and Jonathan Cohen claim that even overt compatibilists turn out to be covert dualist libertarians, with an intuitive concept of free will as 'uncaused causer' of our actions.[36] This claim is not very strongly argued,[37] but according to Greene and Cohen these alleged libertarian intuitions arise because humans have evolved two distinct cognitive subsystems, for 'folk physics' and 'folk psychology'. The folk-psychology subsystem gives us an intuitive concept of agents as uncaused causers of their actions, which is untenable since 'we have little reason to doubt that (i) the state of the universe 10,000 years ago, (ii) the laws of physics and (iii) the outcomes of random

[34] P. Shaw et al., 'Intellectual Ability and Cortical Development in Children and Adolescents', *Nature* 440.7084 (2006), pp. 676–79.

[35] Glannon, *Brain, Body and Mind*, p. 49.

[36] Greene and Cohen, 'For the Law, Neuroscience Changes Nothing and Everything'.

[37] The grounds offered in support of it are quotations from two publications on psychology and criminal law, whose authors apparently slip into a dualist distinction between people and their brains, and an appeal to readers' intuitions about the imaginary case of 'Mr Puppet', whose genetics and personal history have been carefully designed by a villainous group of scientists to ensure that he becomes a psychopathic criminal.

quantum mechanical events are together sufficient to determine everything that happens nowadays, including our own actions'.[38] While neuroscience adds nothing conceptually to the argument for hard determinism, Greene and Cohen think it can provide a publicly compelling demonstration of the argument's force. It does this by turning the mind from a 'black box' into a 'transparent bottleneck': as we learn more about the mechanistic brain processes resulting in decisions, it becomes increasingly plain that the uncaused-causer kind of free will is an illusion.[39]

Compatibilist critics emphasize what Greene and Cohen acknowledge: that neuroscience in itself adds nothing new to familiar arguments about compatibilism and incompatibilism.[40] But, they maintain, it is simply mistaken to believe that genuine free will needs radical indeterminacy or requires agents who are uncaused causers of their actions. As Walter Glannon puts it, 'what matters in having or lacking free will is not so much whether external *sources* influence the mental states issuing in our action; indeed, they do. What matters more is whether we have the internal *resources* to identify with our desires, beliefs, and intentions, and execute them in actions'.[41] This requirement has a negative aspect and, in some accounts, a positive one as well. The negative requirement is Hume's 'liberty of spontaneity': that we are not prevented by coercion, compulsion or constraint from acting on our beliefs, desires and intentions. The positive requirement is that we should be able to own those beliefs, desires and intentions as ours. For example, in Harry Frankfurt's model of first- and second-order desires, persons are capable of forming first-order desires to do certain things, and second-order desires to have certain first-order desires. Free willing and action occur when an effective first-order desire is aligned with a second-order desire, and the person identifies with both.[42] Provided mental states can be causally efficacious, the fact that the 'mental springs'[43] of our actions are the products of physical causal processes does not stop us making them our own.

[38] Greene and Cohen, 'For the Law, Neuroscience Changes Nothing and Everything', pp. 244–45.

[39] Ibid., p. 246.

[40] See, for example, Michael S. Moore, 'Libet's Challenge(s) to Responsible Agency', in Sinnott-Armstrong and Nadel (eds.), *Conscious Will and Responsibility*, pp. 207–30 (218).

[41] Glannon, *Brain, Body and Mind*, pp. 49–50.

[42] Harry S. Frankfurt, 'Freedom of the Will and the Concept of a Person', *Journal of Philosophy* 68.1 (1971), pp. 5–20.

[43] Glannon, *Brain, Body and Mind*, p. 49.

The problem of 'shrinking agency'[44]

Nadelhoffer calls Libet and Wegner's challenge a 'global agential threat': it raises the possibility that there might be no such thing as conscious, willed agency. As we saw in the previous section, the philosophical literature offers various ways of resisting that threat. However, even if that global threat can be seen off, Nadelhoffer believes other areas of psychological research pose 'partial agential threats' by suggesting the conscious mind has a much less significant role in behaviour than we usually think.[45] '[W]hile this view admittedly leaves room for conscious agency,' he remarks, 'it does not leave nearly as much as most libertarians and compatibilists would presumably prefer.'[46]

Automaticity is one such threat. The role of automatic processing in moral judgement was discussed in the last chapter, but according to John Bargh and his colleagues, it can also be found in almost all psychological phenomena.[47] Many preconscious automatic processes may influence subsequent conscious behaviours. One example is the unconscious mimicry of others' behaviour, which can reinforce social bonding and group cohesion. Face perception is another: unconscious reactions to a person's face can influence judgements about their trustworthiness and competence, and even sway voting behaviour in elections. In a different kind of automaticity, labelled 'post-conscious' or 'goal-dependent' by Bargh et al., conscious intentions or decisions initiate mental activities that require minimal conscious control. For example, automatic processing plays an important role in decision-making, particularly where numerous complex factors have to be weighed up. Automatic mental processes may have evolutionary origins: many can be observed even in very young children, and some are found to track cognitive tasks that could have been important for group cohesion in human evolutionary history.

Why might automaticity pose even a partial 'agential threat'? Perhaps because it suggests we are less in control of our own actions than we usually

44 The phrase is taken from Nadelhoffer, 'The Threat of Shrinking Agency'.

45 Ibid., p. 178.

46 Ibid., p. 180.

47 John A. Bargh et al., 'Automaticity in Social-Cognitive Processes', *Trends in Cognitive Sciences* 16.12 (2012), pp. 593–605 (601). The following examples are all taken from this source.

imagine, and this undermines our self-understanding as agents: '[a]fter all, the less conscious *willing* we are able to do, the less free *will* we are able to have'.[48] A related worry is that our behaviour may be influenced by all kinds of factors of which we are unaware and which we do not control, rendering us vulnerable to manipulation by anyone from advertisers after our money to politicians after our votes.[49]

A second phenomenon that could deflate our self-image as conscious agents is *ego depletion*, in which the use of self-control in one situation diminishes subjects' capacity to exercise it again in a subsequent task. This was described in the late 1990s by Roy Baumeister and his colleagues, who used it as the basis for a so-called 'resource model' of self-control.[50] In this model, self-control is something like an energy store, and if it is used in a task that requires resistance to temptation or suppression of emotions, the resource is depleted. Baumeister et al. also suggest that it can be increased through training, though they admit that attempts to do this have had mixed success.[51] In Nadelhoffer's view, the phenomenon of ego depletion emphasizes the fragility of conscious agency and control.[52]

Although many studies have supported Baumeister's findings and model of self-control, both the model and the phenomenon of ego depletion have increasingly been questioned in recent years.[53] Some authors have suggested the phenomenon may be partly or wholly produced by statistical artefacts,[54] or by subjects' own implicit theories about self-control.[55] A recent large-scale study failed to find statistically significant evidence for it, though Baumeister and his colleague Kathleen Vohs have disputed the relevance of that study's

48 Nadelhoffer, 'The Threat of Shrinking Agency', p. 178.

49 Cf. Andrea Kiesel et al., 'Unconscious Manipulation of Free Choice in Humans', *Consciousness and Cognition* 15 (2006), pp. 397–408.

50 Roy F. Baumeister et al., 'Ego Depletion: Is the Active Self a Limited Resource?' *Journal of Personality and Social Psychology* 74.5 (1998), pp. 1252–65; Roy F. Baumeister, Kathleen D. Vohs and Dianne M. Tice, 'The Strength Model of Self-Control', *Current Directions in Psychological Science* 16.6 (2007), pp. 351–55.

51 Baumeister et al., 'The Strength Model of Self-Control', p. 354.

52 Nadelhoffer, 'The Threat of Shrinking Agency', p. 181.

53 See Michael Inzlicht and Elliot Berkman, 'Six Questions for the Resource Model of Control (and Some Answers)', *Social and Personality Psychology Compass* 9.10 (2015), pp. 511–24.

54 Evan C. Carter and Michael E. McCullough, 'Publication Bias and the Limited Strength Model of Self-control: Has the Evidence for Ego Depletion Been Overestimated?' *Frontiers in Psychology* 5.823 (2014), pp. 1–11.

55 Veronika Job, Carol S. Dweck and Gregory M. Walton, 'Ego Depletion – Is It All in Your Head? Implicit Theories about Willpower Affect Self-Regulation', *Psychological Science* 21.11 (2010), pp. 1686–93.

design to the phenomenon of ego depletion.[56] It seems that arguments about the reality of ego depletion are not over just yet.

Addiction and autonomy

Addiction is one issue on which questions about agency and self-control clearly have a bearing.[57] It is a particularly challenging topic for accounts of free will and responsibility, because addictions seem to diminish volitional control of behaviour, but do not obliterate it. Indeed, obtaining and using addictive substances may require complex and carefully planned activity.

The neurobiology of addiction is not thoroughly understood, but in an influential model proposed by Steven Hyman, addictive substances usurp the dopamine systems of the brain.[58] The neurotransmitter dopamine is thought to function as a learning signal, shaping and reinforcing patterns of behaviour that will maximize future success in obtaining rewards like food, safety or sex. Successful reward-seeking behaviours tend to be repeated and may become automatic, triggered by predictive cues. Dopamine is not a signal for the *pleasure* of rewards but rather their 'incentive salience': not 'liking' them but 'wanting' them.[59] Different levels and patterns of dopamine release indicate that things are as expected, better or worse in terms of the gaining of rewards.

Addictive drugs work by various mechanisms, but all increase dopamine release and thereby mimic this natural reward signal. Unlike natural rewards, however, they have no value in terms of homeostasis, survival or reproduction. Because they act directly on dopamine levels in the brain, they produce a

[56] Martin S. Hagger and Nikos L. D. Chatzisarantis, 'A Multilab Preregistered Replication of the Ego-Depletion Effect', *Perspectives on Psychological Science* 11.4 (2016), pp. 546–73; Roy F. Baumeister and Kathleen D. Vohs, 'Misguided Effort with Elusive Implications', *Perspectives on Psychological Science* 11.4 (2016), pp. 574–75.

[57] In the 1960s the World Health Organization sought to clarify the terminology in this area by replacing the term 'addiction' with 'drug dependence' in its official publications: WHO Expert Committee on Addiction-Producing Drugs, *Thirteenth Report* (World Health Organization Technical Report Series, 273, Geneva: World Health Organization, 1964), pp. 9–10, online at http://www.who.int/medicines/areas/quality_safety/reports/en/ (accessed 8 March 2016). However, 'addiction' is still widely used in both academic and general literature, and is the term I shall generally use.

[58] Steven E. Hyman, 'The Neurobiology of Addiction: Implications for Voluntary Control of Behavior', *American Journal of Bioethics* 7.1 (2007), pp. 8–11; Hyman, 'The Neurobiology of Addiction: Implications for Voluntary Control of Behavior', in Illes and Sahakian (eds.), *The Oxford Handbook of Neuroethics*, pp. 203–17.

[59] Kent C. Berridge and Terry E. Robinson, 'What Is the Role of Dopamine in Reward: Hedonic Impact, Reward Learning, or Incentive Salience?' *Brain Research Reviews* 28.3 (1998), pp. 309–69.

reward signal that is unresponsive to experience, even unpleasant experiences caused by the drug-taking (such as a smoker's cough). This reward signal can lead addicted individuals to value drug use more highly than other goals (sometimes even health, work and relationships), and can powerfully shape their behaviour towards obtaining and using drugs. According to Hyman, this model explains effects such as the impairment of decision-making and voluntary control in addiction, and its long-term persistence.[60]

Hyman alludes to a question frequently asked about addiction: should it be regarded as a disease, a moral condition or both?[61] He favours a disease model, while still wishing to hold addicted people responsible for their behaviour; but his account is radically challenged by Bennett Foddy and Julian Savulescu, who argue that addiction is *neither* a disease *nor* a moral condition.[62] Addictive desires, in their view, are merely 'pleasure-oriented desires' no different in kind from others like the desire for strawberries – though they differ from most such desires in their peculiar strength, their origins in particular social contexts and the social stigma attached to them.[63] Foddy and Savulescu deny the normative significance of Hyman's distinction between adaptive natural reward signals and non-adaptive drug-induced ones: people choose all kinds of non-adaptive goods that may even threaten their survival. Pleasure is an 'intrinsic good';[64] we must allow that rational autonomous individuals 'might simply value the pleasure of heroin use above the value of their health or their job', or even regard their drugs as 'more important to them than food'.[65] These may be harmful or risky choices, but many people make harmful or risky choices in pursuit of goods that they value, and we generally respect those choices as autonomous.

60 Hyman, 'The Neurobiology of Addiction' (2011), pp. 211–14.

61 Hyman, 'The Neurobiology of Addiction' (2007), p. 8.

62 Bennett Foddy and Julian Savulescu, 'Addiction Is Not an Affliction: Addictive Desires Are Merely Pleasure-Oriented Desires', *American Journal of Bioethics* 7.1 (2007), pp. 29–32; see also idem, 'Addiction and Autonomy: Can Addicted People Consent to the Prescription of Their Drug of Addiction?' *Bioethics* 20.1 (2006), pp. 1–15. For different and less radical critiques of Hyman see, for example, Louis C. Charland, 'Affective Neuroscience and Addiction', *American Journal of Bioethics* 7.1 (2007), pp. 20–21; Neil Levy, 'The Social: A Missing Term in the Debate over Addiction and Voluntary Control', *American Journal of Bioethics* 7.1 (2007), pp. 35–36.

63 Foddy and Savulescu, 'Addiction Is Not an Affliction', pp. 29–30.

64 Ibid., p. 30. In this article they ignore Hyman's point that dopamine does not signal pleasure but incentive salience, though in their earlier essay 'Addiction and Autonomy' they make the distinction a little more clearly.

65 Ibid., pp. 31, 32. They also argue on the strength of published evidence that such situations are much rarer than popularly supposed.

Foddy and Savulescu's argument might seem persuasive *if* one refuses to attach normative significance to the natural and denies any objective ordering of human goods. As they suggest, Hyman's view – that natural reward mechanisms are 'usurped' by addictive drugs – relies on intuitions that adaptive behaviours have particular normative significance and (some) human goods can be objectively ranked. If those intuitions are true, it looks much more plausible that someone who prefers heroin to food is just terribly mistaken about where his or her good really lies.[66]

How might one make sense of such intuitions? There are no doubt various ways to do so: an Aristotelian-Thomist view of natural law would be one; the kind of theological ethic developed in this book is another.[67] Not that either of these will commend itself to Foddy and Savulescu, who conclude that '[m]uch of the disrepute attached to addiction has been illiberal and the result of one group, often the dominant political or religious group, applying their norms for personal living to others, who share a different ideal of the good life'.[68] Perhaps they are right that the kind of liberalism which avoids making substantive claims about the human good cannot support a negative evaluation of addiction. The judgement that something has *gone wrong* with the life of an addicted person might be – in this specific sense – 'illiberal'. If so, this in itself should not deter theological ethicists from making such a judgement, however much they might wish to avoid the more general opprobrium suggested by 'illiberality'.

It might be possible to make a *liberal* case (in the strict sense) that Foddy and Savulescu are wrong, if one could show that addicted individuals' autonomy is significantly impaired. Foddy and Savulescu deny that it is, if autonomy is understood in a 'thin' way as 'a competent person, in possession of the relevant facts making a free decision about what to do

[66] It is of course a further question what practical conclusions should follow from this view: Hyman does not believe it supports compulsory treatment of addictions, though he thinks it might justify a measure of coercion: 'The Neurobiology of Addiction' (2011), p. 213.

[67] Alert readers might wonder if there is a contradiction here with the previous chapter, where I argued that a theological ethic should regard human intuitions about the good with suspicion. As I also argued there, however, suspicion does not mean rejection out of hand, but critical scrutiny – and where justified, appropriation – in the light of the Gospel. In this theological perspective, the *intuition* that a preference for heroin over food reflects a disordering of desire will not do normative work in establishing what a proper ordering of human goods looks like; but it may play a valuable role in drawing attention to this as a question requiring a theological response.

[68] Foddy and Savulescu, 'Addiction Is Not an Affliction', p. 32.

with his or her life'.[69] Yet addicted people often claim that they use drugs against their will – a claim that according to Neil Levy is supported in various ways by experience. Levy therefore finds Foddy and Savulescu's denial implausible.[70] He agrees that addiction is not compulsion, and does not obliterate autonomous choice; but he holds that it does demonstrate a failure to integrate one's desires and volitions into a coherent life-project.[71]

Levy also agrees with Foddy and Savulescu that addictive and normal desires share a common mechanism, which he identifies as the phenomenon of ego depletion discussed earlier. But unlike them he concludes from this that anyone, addicted or not, may have their autonomy impaired by persistent desires.[72] Others have also emphasized the similarities in the way incentive salience operates in both addictive and non-addictive desires, arguing that the discovery of these common mechanisms discloses 'the addict in us all'.[73]

Moral responsibility

If scientific studies raise doubts about whether we have free will as traditionally understood, do they also challenge the notion that we are responsible for our actions? This question could apply to both moral and legal responsibility, though in this chapter I shall focus mostly on moral responsibility.

What is meant by moral responsibility? Much of the concern seems to be for the tracking of praise and blame (and therefore also judgements about what punishment or reward is deserved). The heart of the matter is that it seems unfair to blame me for doing something bad if I had no possibility of choosing *not* to do it. Conversely, I might not deserve praise for a good deed

[69] Bennett Foddy and Julian Savulescu, 'Autonomy, Addiction and the Drive to Pleasure: Designing Drugs and our Biology: A Reply to Neil Levy', *Bioethics* 20.1 (2006), pp. 21–23. They set the bar very high for an influence to count as an impairment of autonomy in an ethically relevant sense, arguing that neither the behavioural urges experienced by a person with OCD, nor a gun held to one's head by an armed robber – let alone addicts' desires for their drugs – counts as 'compulsion': Foddy and Savulescu, 'Addiction and Autonomy', pp. 6–7.

[70] Neil Levy, 'Addiction, Autonomy and Ego-Depletion: A Response to Bennett Foddy and Julian Savulescu', *Bioethics* 20.1 (2006), pp. 16–20.

[71] Neil Levy, 'Autonomy and Addiction', *Canadian Journal of Philosophy* 36 (2006), pp. 427–48. His account has clear debts to Frankfurt's model of first and second-order desires, though Levy also argues that standard philosophical theories of autonomy fail to give satisfactory accounts of the phenomenology of addiction, because they lack the kind of life-history dimension that his own proposal includes.

[72] Levy, 'Addiction, Autonomy and Ego-Depletion'.

[73] Brendan Dill and Richard Holton, 'The Addict in Us All', *Frontiers in Psychiatry* 5.139 (2014), pp. 1–20.

that was not freely chosen. Behind these thoughts (and implicit in much of the discussion about neuroscience and moral responsibility) is the Kantian idea that 'ought' implies 'can'.[74]

In the terms of the traditional free will-determinism debate, libertarians are likely to think that moral responsibility therefore requires some form of causal openness in mental processes. For some this may be supplied by a substance-dualist view of the mind and brain (though they then face the further challenge of accounting for mental causation); for others, quantum indeterminacy in the physical processes of the brain could help to underwrite the necessary causal openness.[75] Taking a hard-determinist position, Joshua Greene and Jonathan Cohen agree that responsibility requires such causal openness, but maintain that a scientific understanding of causality leaves no room for it, quantum indeterminacy notwithstanding: 'If it turns out that your ordering soup is completely determined by the laws of physics, the state of the universe 10,000 years ago, *and* the outcomes of myriad subatomic coin flips, your appetizer is no more freely chosen than before.'[76] They believe that as this becomes more widely known, it will undermine conventional notions of responsibility. They hope that in the legal sphere, this will result in retributivist theories of responsibility and punishment being displaced by consequentialist accounts: 'At this time, the law deals firmly but mercifully with individuals whose behaviour is obviously the product of forces that are ultimately beyond their control. Some days, the law may treat all convicted criminals this way. That is, humanely.'[77]

Compatibilists, on the other hand, have little difficulty thinking that genuine responsibility is possible in a causally closed physical universe, on occasions when agents are not subject to constraints that make it impossible for the causes of their actions to include reasons that they can own.[78]

[74] According to Roger White, Kant does not seem to have used the actual phrase '"ought" implies "can"' in any of his writings, but the idea is central to his moral philosophy – for example in his argument for freedom in the second *Critique*: Roger M. White, '"Ought" Implies "Can": Kant and Luther, a Contrast', in George MacDonald Ross and Tony McWalter (eds.), *Kant and his Influence* (Bristol: Thoemmes, 1990), pp. 1–72.

[75] For example, in Kane's theory of 'self-forming actions': Robert Kane, 'Some Neglected Pathways in the Free Will Labyrinth', in Kane (ed.), *The Oxford Handbook of Free Will*, pp. 407–37.

[76] Greene and Cohen, 'For the Law, Neuroscience Changes Nothing and Everything', p. 237, emphasis original.

[77] Ibid., p. 255.

[78] See, for example, Moore, 'Libet's Challenge(s) to Responsible Agency', pp. 213–19.

Turning to the newer agential challenges identified by psychology and neuroscience, Daniel Wegner believes his theory of apparent mental causation makes only a minor difference to our understanding of responsibility.[79] In his account, the experience of conscious will is best understood as 'the mind's way of signaling that it might have been involved in causing the action'.[80] This mental signal marks certain events as our actions, and prompts us to feel the moral emotions – such as pride or guilt – appropriate to those actions. This is an illusion, but a good and necessary one that keeps us functioning: 'The feeling of doing is how it seems, not what it is – but that is as it should be. All is well because the illusion keeps us human.'[81] But it is far from clear that this makes so little difference: as Nadelhoffer argues, if our sense of being in conscious control of our actions turns out to be an illusion, this is likely to have much further-reaching implications than Wegner allows:

> By leaving 'us' out of the story about agency and action, it removes us from the sphere of moral desert altogether … if you, like me, don't think it makes sense to say that people deserve blame (or praise) for things over which they have no ultimate control, then [Wegner's] epiphenomenalism represents the greatest threat of all to free will and desert-based responsibility.[82]

Note that Nadelhoffer here echoes two widely shared assumptions we have already encountered: that moral responsibility has to do primarily with tracking desert of praise and blame, and that to be morally responsible we must have 'ultimate control' over our actions. As we shall see, a theological analysis might complicate these assumptions, to say the least.

The first part of this chapter has surveyed some of the extensive debates about free will, autonomous agency and moral responsibility provoked by neuroscientific and social-psychological studies of volition and related topics. We have seen various ways in which the studies of Libet and other neuroscientists, together with psychological research on phenomena like automaticity, ego depletion and addiction, have problematized the view that human beings are autonomous agents in conscious control of, and responsible for, their actions. But we have also seen how vigorously contested almost every

79 Wegner, *The Illusion of Conscious Will*, p. 334.
80 Ibid., p. 336.
81 Ibid., p. 342.
82 Nadelhoffer, 'The Threat of Shrinking Agency', p. 183.

aspect of these arguments is, from the design of the experiments through to the practical moral conclusions they are used to support. Along the way, we have also noted various important assumptions that often go unquestioned. As we turn to theological analysis of these issues, we shall find that some of the issues raised by neuroscience and philosophy must be reframed, and premises often taken for granted may be called into question.

Theological perspectives

That theological engagement will proceed in three stages. First, I shall revisit the question of mental causation and concerns about the 'global agential threat' posed by some scientific accounts. I shall argue that the Christian doctrine of providence creates space for an understanding of causation rich enough to allow for genuine mental causation and agency. Second, I shall relocate questions about determinism, free will and responsibility within an Augustinian account of freedom and sin. I shall suggest that this account challenges some common assumptions about what genuine freedom and responsibility require. As a concrete example, I shall revisit some of the questions discussed earlier about addiction, freedom and responsibility – this time in dialogue with the psychiatrist and theologian Christopher Cook. The discussion of addiction will extend into the final subsection, in which – again in dialogue with Cook – I shall briefly consider treatment and recovery in the light of Christian themes of grace and salvation.

Determinism, causation and providence

At the heart of the 'global agential threat' that Nadelhoffer finds in the work of Libet and Wegner is the question of whether mental events, including conscious willing, can cause physical events such as bodily movements. If not, the notion of conscious will might indeed prove to be an illusion, with all the troubling implications for our self-understanding as free and responsible agents discussed earlier. How might Christian theology respond to this question?

We can gain some purchase on it by revisiting Nancey Murphy's dispute with Jaegwon Kim. Kim's reductionist denial of mental causation relies on a univocal view of causation, in which the existence of one true and complete causal explanation excludes the possibility of additional true causal explanations of the same event. Hence, he will not countenance the possibility that there can be both physical and mental causes of the same physical event (such as a neuron firing in the brain), since this would amount to a causal over-determination of the event, which he rejects.[83] Much of the discussion of neuroscience and free will, to the extent that it is framed by a natural scientific approach that limits itself to the investigation of efficient causes, proceeds implicitly or explicitly on the same assumption. If one complete (efficient) causal explanation of an event in the brain can be found (it is thought), other explanations are thereby excluded.

By contrast Christian thinkers – seeking to do justice to divine sovereignty, human creatures' freedom and responsibility, and the physical necessities to which those creatures are subject – have often maintained richer and more complex understandings of causation, in which more than one true causal story can be told about the same events. For example, they have often been ready to make use of an Aristotelian view in which material, formal, efficient and final causation are different causal schemata, not fully translatable into one another, and causal explanations in terms of each can be simultaneously true of the same effect.[84]

In a very different theological voice, Karl Barth – profoundly suspicious of the Christian use of Aristotelian metaphysics – observes that early Lutheran and Reformed dogmaticians were able to affirm God's governance of the world against both 'the Stoic doctrine of fate' and 'the Epicurean doctrine of chance'.[85] It may not be entirely wayward to see in this opposition of 'fate' and 'chance' a rough counterpart to modern oppositions between determinism and the kind of radical indeterminacy that some libertarians think freewill requires. If a doctrine of divine governance could steer a middle course between *fatum* and *fortuna*, perhaps it also has the resources to negotiate present-day disputes between determinists and libertarians.

According to Barth, God's governance of the world transcends and uses both creaturely freedom and necessity, and there is no contradiction between divine sovereignty and creaturely freedom: 'The freedom of [the creature's] activity does not exclude but includes the fact that it is controlled by God. It is God who

[83] Kim, 'Blocking Causal Drainage', pp. 160–64.

[84] See Simon Oliver, 'Aquinas and Aristotle's Teleology', *Nova et Vetera* 11.3 (2013), pp. 849–70.

[85] Karl Barth, *Church Dogmatics* (G. W. Bromiley and T. F. Torrance (ed.), 13 vols., Edinburgh: T&T Clark, 1956–75), vol. 3.3, p. 162.

limited it by law and necessity and it is God who created it free.'[86] Creaturely freedom, says Barth, operates on the basis, and within the framework and limits, of divine permission. This limitation is not a compulsion laid on the creature, but rather a necessary *condition* of genuine creaturely freedom: the attempt to act outside the limit of divine permission would be a self-destructive effort to claim godlike freedom. Barth would therefore roundly reject the notion that genuine (creaturely) freedom requires us to be the uncaused causers of our actions. Moreover, he continues, 'all creaturely activity aims at a certain effect', but the goal and the outcome are beyond the creature's control:

> Whether the effect comes, and if so how it comes, is a completely new factor in relation to the activity. This is true whether we consider it from the standpoint of necessity or from that of freedom. And if it is God who controls creaturely occurrence and not fate or chance, then we have to say quite baldly that the decisive moment, the very meaning of creaturely activity, its effect, and the goal or end in which it culminates, are all the gift and dispensation of God.[87]

In other words, every aspect of a creature's activity – including the material conditions that make it possible, the physical cause and effect that it involves and the goal to which it is directed – is governed and given space by divine providence.

Barth is not offering a theory of causality, but expressing what (he thinks) Christian theology is committed to saying about the divine governance of creation and creaturely freedom. In so doing, he establishes some boundaries and parameters for what a philosophical theory of mental causation must say, if it is to articulate this Christian vision of creaturely freedom truthfully. Theology as such need not – and probably should not – be committed to any particular philosophical perspective, but it can be hospitable to any philosophical account that helps it articulate this vision. Murphy's theory of downward causation, for example, seems to fit well with Barth's theological account, and could have a valuable contribution to make in filling out the outlines of his biblically and theologically shaped picture.

This view of mental causation does not depend on a particular understanding of the soul or its relationship to the body. Rather, it keeps open a space within

[86] Ibid., p. 166.
[87] Ibid.

which even a physicalist view of brain and mind does not entail the hard-determinist view that there is no true freedom or moral responsibility. In other words, Christians seeking to think Christianly about determinism and free will in the context of neuroscience need not be constrained by a naturalist framing of the problem. The physical laws by which brains operate, the particular structures and workings of an individual's brain, the experiences to which that individual responds and the goal that she seeks in making a decision 'are all the gift and dispensation of God'. That being the case, a cause-and-effect explanation of the physical processes occurring in her brain as she makes her decision is not a zero-sum alternative to an account of her reasons, or a teleological explanation in terms of the goods and goals to which her decision is directed.

Neuroscience, responsibility and sin: What kind of freedom?

Earlier in the chapter, we saw that neuroscientific research has raised two distinct questions about free will: the question of mental causation, which I discussed theologically in the last section, and the more familiar one about the (in) compatibility of causal determinism with free will. On the latter, incompatibilists may argue (like Greene and Cohen) that neuroscience supports hard determinism, even if it adds nothing of substance to the philosophical arguments. Compatibilists might reply (like Michael Moore) that it is philosophically confused to believe that neuroscience challenges compatibilism. I suggested in the chapter introduction that incompatibilists and compatibilists often look for different kinds of freedom: libertarians might think something like Hume's 'liberty of indifference' is needed for genuine free will (and hard determinists would say it is impossible in a causally determined world) whereas for compatibilists the 'liberty of spontaneity' – freedom from coercion, compulsion and constraint – suffices. Some theologians also adopt one or other of these positions.[88]

As also noted in the Introduction, it is widely believed that free will requires the possibility of choice between alternative options. Yet the Christian tradition

[88] For critiques of compatibilism and arguments that Christians should be libertarians, see, for example, Marcel Sarot, 'Christian Faith, Free Will and Neuroscience', in Peter Jonkers and Marcel Sarot (eds.), *Embodied Religion: Proceedings of the 2012 Conference of the European Society for Philosophy of Religion* (Utrecht: Igitur, 2013), pp. 105–20; Alan Torrance, 'Developments in Neuroscience and Human Freedom: Some Theological and Philosophical Questions', *Science and Christian Belief* 16.2 (2004), pp. 123–37. For a Christian defence of compatibilism, see, for example, Aku Visala, 'Theism, Compatibilism and Neurodeterminism: A Response to Marcel Sarot', in Jonkers and Sarot (eds.), *Embodied Religion*, pp. 121–39.

has sometimes thought of free will in rather different ways from either the liberty of spontaneity or the liberty of indifference: ways that also question whether it really requires alternative possibilities. A well-known passage from Augustine's *City of God* illustrates this:

> Neither are we to suppose that because sin shall have no power to delight [the saints in heaven], free will must be withdrawn. It will, on the contrary, be all the more truly free, because set free from delight in sinning to take unfailing delight in not sinning. For the first freedom of will which man received when he was created upright consisted in an ability not to sin, but also in an ability to sin; whereas this last freedom of will shall be superior, inasmuch as it *shall not be able to sin*.[89]

Augustine believes that the ability to choose between different possibilities (specifically, to sin or not to sin) does play some part in the human experience of freedom. This was (he thinks) true in particular of Adam and Eve before the Fall, for whom it was possible not to sin (*posse non peccare*) but also possible to sin. However, this form of freedom is at best a preparatory state: as the quotation above shows, for Augustine, the fullest and highest freedom consists in the inability to sin (*non posse peccare*).[90] He holds that even if humans had not sinned, their freedom to choose not to sin would have been displaced by this higher freedom when they were perfected in glory.[91] For Augustine, then, the highest and most perfect freedom does not require the possibility of alternative choices, but consists in the ability to bind one's will indissolubly to the good. This understanding of freedom is closer to compatibilist philosophical accounts, which emphasize the ability to act for reasons one can own, than to libertarian views. One could perhaps call it compatibilism of a sort, as Jesse Couenhoven does.[92] However (as he acknowledges), it is a rather

[89] Augustine, *The City of God* (Philip Schaff (ed.); Marcus Dods (trans.); *Nicene and Post-Nicene Fathers*, 1–02; Edinburgh: T&T Clark, 1886), 22.30, emphasis added; online at http://www.ccel.org/ccel/schaff/npnf102.i.html (accessed 30 March 2016).

[90] As Couenhoven shows, Augustine makes this case – particularly in his later anti-Pelagian writings – with reference not only to the saints in heaven but also to the sinless life of Jesus and the freedom of the triune God: Jesse Couenhoven, 'The Necessities of Perfect Freedom', *International Journal of Systematic Theology* 14.4 (2012), pp. 396–419.

[91] See Ian A. McFarland, '"Willing Is Not Choosing": Some Anthropological Implications of Dyothelite Christology', *International Journal of Systematic Theology* 9.1 (2007), pp. 3–23 (3 n. 1).

[92] Jesse Couenhoven, *Stricken by Sin, Cured by Christ: Agency, Necessity, and Culpability in Augustinian Theology* (Oxford: Oxford University Press, 2013), esp. ch. 5. There are some differences between Couenhoven's rich, detailed and insightful account and mine, a few of which will be noted in what follows; but the general direction of my argument is similar to his.

different compatibilism from the philosophical versions that we met earlier in the chapter, and it will be important to differentiate this Augustinian view of freedom from those forms of compatibilism as well as the libertarian views I surveyed earlier.[93]

Our greatest freedom consists in the ability to bind our wills to the good. In the present age, though, our universal human condition is the opposite of that state: our wills are bound not to the good but to sin, so that it is impossible for us to avoid sinning (*non posse non peccare*). This is the condition that Augustine attempts to describe and explain by means of his doctrine of original sin. As Couenhoven has argued, Augustine's account of original sin has several distinct components, which are separable from each other and do not necessarily all stand or fall together.[94] Some are problematic for various reasons: one is the link between the sexual generation of offspring and the transmission of original sin (which in Couenhoven's judgement Augustine never convincingly explains[95]); another is the claim that original sin began with the historical fall of a first human couple.[96] However, these problems need not undermine the central insight of Augustine's doctrine, which is that our present condition of being "bound to sin" is inherited by each of us before we are ever in a position to choose it.[97] We are born into lives, relationships and social structures distorted in all kinds of ways by sin. These distortions mis-shape our beliefs, desires, affections and wills so that as soon as we are capable of willing and acting ourselves, we do so in ways distorted by sin. Thus, we in turn contribute our willed and chosen sins to the condition of sin we have inherited. As Alistair McFadyen emphasizes, original sin so understood does not obliterate our capacity for willing and choosing, but rather co-opts and diverts it.[98]

93 Couenhoven opposes his 'Augustinian compatibilism' to the libertarianism which he regards as the dominant present-day view (e.g., *Stricken by Sin*, pp. 8–9). However, as we have seen, it is not clear that libertarianism *is* the dominant view in the discussions with which this chapter is concerned, which makes it all the more important to differentiate a theological account from philosophical compatibilism.

94 Couenhoven, *Stricken by Sin*, ch. 1.

95 Ibid., pp. 42–45.

96 For an account of how an Augustinian doctrine of original sin might be understood, without an historical fall, in the context of human evolution, see Neil Messer, *Selfish Genes and Christian Ethics: Theological and Ethical Reflections on Evolutionary Biology* (London: SCM Press, 2007), pp. 184–95. I argue there that a doctrine of original sin, so understood, offers important insights into some of the darker aspects of our species' evolutionary heritage.

97 Couenhoven, *Stricken by Sin*, ch. 2; so also Alistair McFadyen, *Bound to Sin: Abuse, Holocaust and the Christian Doctrine of Sin* (Cambridge: Cambridge University Press, 2000), pp. 16–18.

98 McFadyen, *Bound to Sin*, ch. 8.

In this Augustinian account, freedom does not depend on the ability to choose between alternative possibilities; nor does responsibility depend on the availability of alternatives or the ability to choose them. In this view, we are responsible or accountable to God for our present state of sinfulness, even though we inherited it before we could ever choose it, and are unable to avoid it.

To many modern minds, this might seem unjust, if not incoherent. It is a deeply held intuition that it is unfair to hold someone responsible for a state of affairs that could not have been otherwise. And the Kantian slogan, '"ought" implies "can"', presses the objection further, suggesting that it is not only unfair but nonsensical to speak of obligation or responsibility in the absence of choice. It is implicit in the meaning of the claim, 'we ought not to sin', that it is possible for us not to sin. Nor is it only modern sensibilities that are offended by the Augustinian view. In Augustine's controversies with the Pelagians – so far as we can judge from the surviving evidence – they insisted that freedom understood as 'a neutral suspension between different possibilities' was both a reality and a condition for moral responsibility.[99] Yet Couenhoven argues that the Augustinian view makes better sense of much human experience: for example, we are accustomed to praise or blame people for qualities of character, and the acts that flow from them, which were not, or not fully, in their power to choose.[100]

What really seems to be going on here is a confrontation between contrasting moral visions, one that has occurred repeatedly with variations in the history of Western Christian thought. In one vision, which could be called Pelagian, questions of praise and blame are foregrounded. A central concern is the tracking of moral accountability so that praise, blame, punishment and reward can be justly apportioned.[101] In the other, which could be called Augustinian,

[99] Ibid., p. 168.

[100] Couenhoven, *Stricken by Sin*, ch. 6.

[101] McFadyen (*Bound to Sin*, pp. 19–22) calls this a 'moral' frame of reference, and complains that modern understandings of sin are often confined to such a moral frame. Part of his complaint is that to do so reflects what he calls 'pragmatic atheism': whether or not we believe in God, that belief makes no practical difference to the ways in which we understand human affairs or act in the world. In opposition to such a pragmatically atheist understanding, he seeks to develop a thoroughly theological account of sin – Augustinian, at least in part – in which moral categories such as praise and blame play a much less prominent part in describing our condition before God. By contrast, Couenhoven's version of an Augustinian doctrine of sin maintains closer links between sin, responsibility, blame and punishment, making use (as noted above) of an Augustinian account of responsibility that does not depend on choice (*Stricken by Sin*, ch. 6). Couenhoven takes McFadyen

praise- and blameworthiness are not absent from the picture, but they are no longer the most important or interesting questions. They should be raised only *after* others have been asked.[102] In this moral vision, the central question could be expressed as: In what state or condition do I stand before God? As Roger White has shown, a millennium after Augustine a similar contest of moral visions was played out in Luther's dispute with Erasmus concerning the freedom of the will.[103] According to White, more recently still, Immanuel Kant took the same side as Erasmus on what was, very roughly speaking, a secularized version of the same issue.

Does 'ought' imply 'can'? On the Augustinian view, not necessarily. In Luther's words, '[i]t is dangerous to believe that the existence of a law implies that it can be obeyed, for the law is fulfilled by the grace of God.'[104] We can still be guilty of sin (not in the modern sense of mere blameworthiness, but in the sense that we stand before God in a sinful condition, which we perpetuate and add to by our own sinful choices and actions) even when it is not in our power *not* to sin. At this point, it is worth emphasizing – perhaps counterintuitively – that this account of sin is good news, because as the quotation from Luther indicates, it is merely the shadow side of the Christian doctrine of grace. In similar vein Karl Barth emphasizes that it is only in the light of God's reconciling work in Christ that we know ourselves to be sinners, and know what sin looks like.[105] I shall return to the theme of grace in the next section.

In the first part of the chapter, we saw how research in psychology and neuroscience appears to pose what Nadelhoffer called 'partial agential threats': suspicions that we have much less free will than we imagine. Nadelhoffer (like many others) thinks anyone who cares about human agency and responsibility is likely to find such threats troubling. However, Christians who hold the

to task for retaining a more conventional understanding of responsibility, which he, Couenhoven, considers a flaw in an account that is strong in many other respects (ibid., pp. 201–2). Couenhoven's account does however introduce some distance between responsibility, blame and punishment, so that blame is not always the appropriate response to sins for which someone is responsible (or there may at any rate be factors that mitigate it), and likewise punishment may not always be the most appropriate response to justly apportioned blame (ibid., ch. 4).

[102] So White, '"Ought" Implies "Can"', p. 71.

[103] Ibid. Key sources for this debate are: Martin Luther, 'Disputation against Scholastic Theology', in *Luther: Early Theological Works* (James Atkinson (trans.); Library of Christian Classics, 16; London: SCM, 1962), pp. 266–73; Erasmus, 'On the Freedom of the Will', in *Luther and Erasmus* (E. Gordon Rupp and A. N. Marlow (trans.); Library of Christian Classics, 17; London: SCM, 1969), pp. 35–97.

[104] Luther, 'Disputation', p. 270.

[105] Barth, *Church Dogmatics*, vol. 4.1, pp. 388–89.

theological view of freedom, responsibility and sin just outlined should be remarkably *un*troubled by them. For one thing, some of the constraints that are said to exist may simply be aspects of our creaturely finitude that we did not know about until now; in Christian perspective, it should never be a cause for regret that as finite creatures, we cannot do everything that we can imagine doing. More importantly, the Christian tradition might not have much of a stake in the kind of freedom called into question by psychological and neuroscientific claims. I have suggested that theology should not be interested primarily in either Hume's 'liberty of indifference' or 'liberty of spontaneity': the freedom that matters most to Christians is the freedom to orient oneself fundamentally to the good. That freedom is compromised not by our creaturely finitude, but by our sinfulness, our fallen condition: the radical alienation from God that we are both born into and perpetuate through our own willing and choosing.[106]

That is not to say, however, that psychology and neuroscience are irrelevant to this account. The development of neural structures, pathways and mechanisms is significantly shaped by a person's experiences and relationships, particularly through childhood and adolescence. Individual variants of brain structure and function can be seen as a physical 'sediment' (to borrow McFadyen's metaphor) of a particular personal history of experiences, relationships and social environment.[107] If aspects of that history are distorted by sin (as they are for all of us, in different ways), we might expect that distortion to leave its traces in the brain. Perhaps, therefore, some of the constraints on freedom mapped by psychology and neuroscience can be understood theologically as ways in which individual, corporate or structural sin compromises our freedom to orient ourselves to the good.

Can this perspective offer any insights into the complex and ambiguous relationship between addiction, freedom and responsibility, discussed earlier in the chapter? In the Augustinian perspective I have outlined, we are all sinners whose wills are bound, not to the good, but to sin; yet as McFadyen emphasizes, this does not obliterate, but rather diverts and co-opts, our willing and choosing. In an important study, Christopher Cook has shown how such

[106] Cf. McFadyen, *Bound to Sin*, pp. 126–30.

[107] McFadyen, *The Call to Personhood: A Christian Theory of the Individual in Social Relationships* (Cambridge: Cambridge University Press, 1990), p. 8 et passim.

an understanding might shed light on alcohol dependence.[108] Like some of the philosophical accounts explored earlier, he draws on Frankfurt's concept of first- and second-order desires to map the complexity of such dependence: one might have a second-order desire to be sober; but that second-order desire is easily overwhelmed by the immediate first-order desire to satisfy the craving for drink.

Cook, however, does not confine himself to the philosophical, but develops an illuminating theological analysis of addiction as sin, drawing on Paul and Augustine. The conflict of desires in an alcohol-dependent person is understood in terms of the divided will or self: 'I do not do the good I want, but the evil I do not want is what I do' (Rom. 7:19). Importantly, he too emphasizes that this is one, perhaps extreme, instance of a common human condition. There may or may not be an 'addict in us all', as suggested by some of the accounts I surveyed earlier, but in this Pauline-Augustinian perspective we are certainly all sinners.[109]

In this theological analysis is addiction a medical or moral condition? Neither, in any simplistic sense, though it can be seen as a bio-psycho-social condition with moral ramifications; but at its most fundamental it is a spiritual predicament, one particular, vivid manifestation of a predicament we all share: a divided self and a bound will. As McFadyen argues, this claim is first and foremost a *theological* one about the condition in which we stand before God.

That said, McFadyen's distinction between theological and moral frames of reference should not be overdrawn. To describe ourselves as sinners is to say that we radically fail to orient our lives to our truest good, and this failure has profound consequences in further failures to realize the good in our own lives and our personal and social relationships – the privation of good, in Augustinian language. As such this theological self-understanding lies at the heart of a Christian ethic: it is among other things a *moral* perspective. McFadyen and Roger White are however right to emphasize that this is a much broader and richer frame of reference than the kind of moral frame in which questions of praise and blame have a central place. If we describe addiction or alcohol dependence in terms of sin, we are saying that they have a moral aspect in which questions of responsibility do arise – but in such a moral vision, questions of praise and blame are by no means the first or most important ones to raise. And

[108] Christopher C. H. Cook, *Alcohol, Addiction and Christian Ethics* (Cambridge: Cambridge University Press, 2006), ch. 6.

[109] Cook, *Alcohol, Addiction and Christian Ethics*, pp. 164–66.

it is particularly important to emphasize, in view of the ways the language of sin has too often been misunderstood and misused, that this perspective gives no warrant for stigmatizing addicted persons or heaping opprobrium upon them. Stigma and opprobrium are simply not what this analysis is about.

Neuroscience, sin and grace

We stand before God in a fallen condition, our wills bound not to the good, but to sin: to a fundamental distortion in our relationship with God, a distortion we both inherit and willingly contribute to, which in its turn spawns distorted relations with one another, ourselves and the created world. We cannot free ourselves by our own efforts from this binding of the will. Yet God's saving work accomplishes for us what we cannot do for ourselves, setting our bound wills free. As I observed earlier, a Christian doctrine of sin is good news, precisely because it is the shadow side of Christian faith in God's grace and salvation. We know about the bondage of the will in the light of what God has done to liberate us by the cross and resurrection of Christ. As McFadyen shows, an Augustinian view of grace, like Augustine's doctrine of sin, can both complicate and enrich the understandings of free will discussed in this chapter.[110] The work of God's grace in us, giving rise to faith (bringing about a fundamental orientation of our being towards God and the good) is a kind of compulsion or coercion. But as McFadyen argues, it is a compulsion that is 'not unwilling, but consensual ... In faith, the will is compelled but, claims Augustine, it is also free. Indeed, it is free precisely because it is compelled in movement towards God and the good'.[111] The compulsion of grace, so understood, does not subvert human willing but 'excites and redirects [it] from within'.[112] This paradoxical Christian vision is powerfully expressed by John Donne:

> Take me to you, imprison me, for I
> Except you enthral me, never shall be free.[113]

[110] See, for example, Augustine, *On the Spirit and the Letter*, in *Anti-Pelagian Writings* (Philip Schaff (ed.); Peter Holmes, Robert Ernest Wallis and Benjamin B. Warfield (trans.); *Nicene and Post-Nicene Fathers*, 1-05; Edinburgh: T&T Clark, 1887), online at http://www.ccel.org/ccel/schaff/npnf105.xi.html (accessed 30 March 2016).

[111] McFadyen, *Bound to Sin*, p. 184.

[112] Ibid., p. 212, emphasis original.

[113] John Donne, 'Batter My Heart, Three-Person'd God', in *Complete Poetical Works* (Delphi Classics, 2012).

If God's grace in Christ does for us what we cannot do for ourselves, what should we make of technological attempts to modify the workings of the brain – such as drugs or surgery – aimed at alleviating what I have characterized as both products and aspects of our sinful condition? Are they to be understood as instruments of God's grace? Or are they attempts to substitute human action for God's saving work: instances of the human pride which, as Barth says, wants to be its own helper and so turns us away from the divine source of our true help?[114] Cook discusses this question in relation to the treatment of alcohol dependence, in a way that demonstrates the complexity of the issue but does not fully resolve it.[115] On the one hand, for example, he sees in the twelve-step programme of Alcoholics Anonymous, with its appeal to a 'Higher Power', a recognition of the need for divine assistance. On the other, he argues that therapeutic techniques and 'anti-craving' drugs 'are grace of a kind, and do produce a form of salvation' – a form, however, not oriented overtly to God but to 'human freedom and fulfilment'.[116]

Perhaps we can gain further clarity here by means of Dietrich Bonhoeffer's distinction between the ultimate and the penultimate.[117] By the 'ultimate', Bonhoeffer means God's last word to humanity, the word of judgement and salvation in Jesus Christ. The 'penultimate' could refer to almost any aspect of life and human activity in this world. In the light of the ultimate, this-worldly human activities stand under God's judgement insofar as we try to save ourselves or reconcile ourselves to God by means of them. Yet by God's grace they can also become *penultimate* activities, preparing the way for the coming of grace into human lives. Bonhoeffer gives examples such as feeding the hungry, housing the homeless and freeing the enslaved.[118] Medicine and health care may likewise have this penultimate character, curing or alleviating conditions that inhibit patients from responding to God's liberating command to 'will to be healthy'.[119] Similar clinical techniques and practices, however, could acquire a very different meaning were they to be used in an attempt to

[114] Karl Barth, *Church Dogmatics*, vol. 4.1, pp. 458–78.

[115] Cook, *Alcohol, Addiction and Christian Ethics*, pp. 186–89.

[116] Ibid., p. 188, citing Doctrine Commission of the Church of England, *The Mystery of Salvation* (London: Church House Publishing, 1997), pp. 31–40.

[117] Dietrich Bonhoeffer, *Ethics* (Ilse Tödt et al. (eds.); Reinhard Krauss, Charles C. West and Douglas W. Stott (trans.); Dietrich Bonhoeffer Works, 6; Minneapolis, MN: Fortress, 2005), pp. 146–70.

[118] Bonhoeffer, *Ethics*, p. 163.

[119] Barth, *Church Dogmatics*, vol. 3.4, pp. 356–74; on medicine as a penultimate activity, see Neil Messer, *Flourishing: Health, Disease, and Bioethics in Theological Perspective* (Grand Rapids, MI: Eerdmans, 2013), pp. 181–82.

transcend human creaturely limits as such: uses of this sort would be more reminiscent of that form of pride which Barth calls '[t]he attempt at self-help'.[120]

When we consider clinical interventions to address what I have identified as physical traces of sin in the brain, the line between responsible penultimate activity and hubristic substitute salvation becomes finer than ever. But the line still exists, and discernment between the two possibilities, though all the more difficult, is crucial. For this reason, while Cook is quite right to discern in therapeutic interventions 'the common grace that pervades creation',[121] it does not seem entirely helpful to describe them as 'a form of salvation ... oriented towards human freedom and fulfilment',[122] because this tends to obscure rather than clarify that line. However, his actual understanding of the place and limits of clinical interventions appears quite close to what I have described as responsible penultimate activity. Such therapies could be considered penultimate insofar as they help people dependent on alcohol to overcome obstacles to living and acting responsibly towards God and neighbour, or open up a space in which they have the opportunity to locate their treatment in the context of questions about their ultimate goods, goals and ends.[123]

Conclusion

In this chapter, I have surveyed a range of questions about free will, responsible agency and autonomy raised by psychology and neuroscience. Neuroscientific studies of willed actions raise familiar questions about determinism and free will, and less familiar ones about mental causation, leading some to claim that conscious will is an illusion. Psychological studies of automaticity and ego depletion may deflate our view of our personal agency. Addiction appears to subvert, without obliterating, autonomy and willed control of behaviour. If consciously willed control of our actions is debunked, deflated or distorted in these ways, that raises questions about our responsibility for those actions.

I have argued in the second part of the chapter that if these questions are relocated in a frame of reference shaped by Christian theological themes of

[120] Barth, *Church Dogmatics*, 4.1, p. 463 et passim.
[121] Cook, *Alcohol, Addiction and Christian Ethics*, p. 187.
[122] Ibid., p. 188.
[123] Cf. ibid., p. 189.

creation, providence, sin and salvation, they can be reframed and addressed in fruitful and interesting ways.

If we understand the world as God's creation, sustained by God's providential care, then creaturely freedom is not the godlike freedom of the uncaused causer, but is both limited and underwritten by God's governance of the world, and both freedom and causal necessity are 'the gift and dispensation of God'.[124] This view of the world calls for a rich understanding of causality in which freedom can coexist with causal determination, and it is hospitable to philosophical accounts (such as Murphy's theory of top-down causation) which can make sense of mental causation of physical effects in the brain and body.

An Augustinian view of sin and grace offers different perspectives on freedom and responsibility from those surveyed in the first part of the chapter. Our greatest and truest freedom consists in the ability to bind our wills indissolubly to the good. This freedom does not require alternative possibilities, and is not adequately described in terms of liberty of spontaneity or liberty of indifference. As sinners and heirs to original sin, we cannot attain such freedom by our own efforts. Our wills are bound to sin, and though we inherited this condition before we chose it, we also contribute to it and pass it on to others through our own willing and choosing. The peculiar mix of freedom and unfreedom seen in addiction can be understood in these terms: as a distortion of human identity and relationship that does not obliterate, but co-opts and subverts, our willing, choosing and acting – and is simply one vivid instance of a human predicament that all of us share.

It is God's grace that liberates our bound wills and gives us the hope of attaining our true freedom. This grace can be described, paradoxically, as a form of compulsion or coercion that sets us free. In the perspective of God's grace, therapeutic interventions addressing particular human predicaments (such as anti-craving drugs or other forms of therapy for addictions) have the potential to be expressions of our 'penultimate' human calling to prepare the way for the coming of God's grace into human lives. In short, the theological themes of providence, sin and salvation have rich potential to complicate this area of neuroethical discussion in illuminating and fruitful ways.

[124] Barth, *Church Dogmatics*, vol. 3.3, p. 166.

5

Consciousness and Its Disorders: Uncle Charlie Revisited

Introduction

Disorders of consciousness such as the permanent vegetative state (PVS) pose some of the most contentious practical bioethical and neuroethical questions. How should patients who appear to have irretrievably lost their capacity for conscious awareness be treated? In particular, is it permissible to withdraw the artificial nutrition and hydration (ANH) required to keep these patients alive? If it is permissible, how should decisions about withdrawing ANH be made in individual cases, and by whom? Such arguments are not confined to the academic literature, but become enshrined in legislation, are played out in the courts and attract high-profile media attention.[1]

In recent years, such debates have been further complicated by functional brain imaging research suggesting that some patients who show all the clinical signs of PVS may actually have some level of conscious awareness. The interpretation of this work has been – and continues to be – contested, with some critics disputing that it really indicates consciousness in these patients at all. But if it does, what are the ethical implications? It is often assumed that if such patients do turn out to be conscious, that will indicate that they are really 'persons' with rights, including the right to life. However, as we shall see later in the chapter, this is disputed by some neuroethicists, who argue that that

[1] For example, the current legal position in the UK was established by the Law Lords' judgement in *Airedale N.H.S. Trust v Bland* (1993) A.C. 789 House of Lords. For one high-profile recent British case, see Sanchia Berg, 'Permanent Vegetative State: A Family's Agony', *BBC News Magazine* (23 September 2016), online at http://www.bbc.co.uk/news/magazine-37444379 (accessed 2 March 2017).

evidence of consciousness in patients with PVS diagnoses might turn out to have more unexpected ethical implications.

Although questions about the treatment of PVS patients, withdrawal of ANH and the like are thoroughly practical, this is one of those neuroethical topics in which the most concrete of practical ethical questions are entangled with the most seemingly abstract of theoretical concepts and debates. What exactly is lost in conditions like PVS, and how is it lost? If neuroimaging studies indicate evidence of 'consciousness', what exactly are they showing evidence of? Neuroscientific studies of consciousness and its disorders are inescapably theory-laden from start to finish: their design, analysis and interpretation will all be informed by philosophical and neuroscientific accounts of consciousness and the self, whether investigators are aware of those influences or not. Ethical arguments about the implications of these studies also depend on theories of consciousness and the self, as we shall see. Therefore, it will be necessary in this chapter to uncover some of these theoretical connections and give at least a brief account of them.

However, the main purpose of this book is to explore what difference it makes to do neuroethics theologically. Christian theologians and ethicists will think differently from their secular colleagues about the conceptual and practical questions raised by disorders of consciousness. If we ask what it means to be a conscious self, Christian theology is likely to raise the question of the soul. Yet among theologians who have engaged with neuroscience, there is wide agreement that any adequately Christian talk of the soul will – at the very least – be radically different from the popular dualistic notions that neuroscience is often supposed to have discredited. The Christian accounts offered by these theologians, which I shall survey and discuss briefly, can critically appropriate insights from neuroscience and the philosophy of mind without much difficulty. However, I shall also suggest that some attempts to respond theologically to neuroscience have troubling implications (possibly unintended by their authors) for the moral and theological status of patients with disorders of consciousness. Theological accounts too may therefore require some careful critical handling and possibly some reappraisal.

Later sections of the chapter return to the practical ethical questions raised by disorders of consciousness. I offer a critical survey of some philosophical arguments about these questions, before exploring how the debates might be reframed theologically. In particular, I argue that the concept of 'person', widely used by all sides in controversies about the treatment of PVS patients, may have less theological-ethical traction than is often supposed. This is where the 'Uncle Charlie' of the chapter title comes in, referring to an early essay by Stanley Hauerwas that makes a similar critical point. Taking up a proposal made by various authors, I offer a different way for Christians to think ethically about these questions, based on a particular reading of the Good Samaritan story (Luke 10:25-37).

Definitions

Before I begin the discussion, some definitions of key terms might be helpful.[2] '*Disorder of consciousness*' (DoC) is a generic term for a condition that occurs when areas of the brain involved in consciousness are damaged or destroyed, for example, by a traumatic injury or a stroke. If the damage is severe enough, it will cause brain death. Otherwise the patient may enter a coma, a completely unconscious and unresponsive state. This is typically time-limited, and some patients progress from it into a *vegetative state* (VS). VS is often described as 'awake but unaware': patients have sleep/wake cycles and can sustain the autonomic functions necessary for life (like breathing), but are unresponsive to stimuli and show no behavioural signs of conscious awareness. If this state lasts long enough, it is described as *permanent* (PVS) and recovery is judged practically impossible.[3] But before that point, some patients progress to a *minimally conscious state* (MCS) in which they show definite signs of awareness and consciousness, though these may be episodic and intermittent. Some patients remain in this state, while others recover higher levels of conscious functioning. The *locked-in state* (LIS) should also be mentioned, though this is not really a disorder of consciousness: locked-in patients have relatively high levels of conscious awareness but are near-totally paralysed, so they cannot communicate by speech or movement. They may however be able to answer questions or spell out words, for example by blinking.[4]

2 For the following, see Joseph J. Fins, 'Neuroethics and Neuroimaging: Moving toward Transparency', *American Journal of Bioethics* 8.9 (2008), pp. 46–52 (47–48).

3 How long depends on the source of the damage. A vegetative state lasting longer than a month is described as 'persistent'. After three months (if the brain damage was caused by oxygen starvation, for example as a result of a cardiac arrest or a stroke) or twelve months (if it was caused by a traumatic injury) it is considered permanent.

4 The classic patient narrative of LIS is Jean-Dominique Bauby's remarkable book *The Diving Bell and the Butterfly* (Jeremy Leggatt (trans.); London: Fourth Estate, 1997).

The borders of consciousness: Neuroimaging studies

In 2006, Adrian Owen and his colleagues reported that they had used fMRI to detect awareness in a patient in the vegetative state. The patient was totally unable to communicate and met all the behavioural criteria for a VS diagnosis. In the study, she was instructed to perform two different mental tasks: to imagine playing tennis, and to imagine visiting all the rooms of her house. The two instructions elicited different patterns of brain activity, indistinguishable from those of healthy volunteers performing the same tasks.[5]

This report built on several years' previous neuroimaging studies indicating various forms of brain activity in patients with disorders of consciousness.[6] These earlier studies were claimed as evidence of conscious awareness even in some patients who appeared to be in VS or MCS. However, their interpretation was ambiguous, because it was possible that most or all of the observed brain activity resulted from automatic processing in the absence of conscious awareness or a sense of self.[7] The study by Owen et al., on the other hand, seemed to give clearer and more direct evidence of conscious awareness, less easily explained as the product of automatic processing.[8]

The potential significance of this report was obvious. It challenged a core assumption built into the understanding of the vegetative state since it was defined in 1972, namely, that VS patients are awake but unaware.[9] If correct, it would have important implications for diagnosis, prognosis and clinical care of patients with DoCs: diagnosis, because it suggested that behavioural signs alone could not reliably indicate that a patient lacked conscious awareness; prognosis, because neural signs of awareness might be followed by behavioural signs, in which case the neural markers could be early indicators of recovery;[10]

[5] Adrian M. Owen et al., 'Detecting Awareness in the Vegetative State', *Science* 313.5792 (2006), p. 1402.

[6] For a survey of some of these, see Fins, 'Neuroethics and Neuroimaging', pp. 48–49.

[7] So, for example, Neil Levy, 'Going beyond the Evidence', *American Journal of Bioethics* 8.9 (2008), pp. 19–21, drawing on the psychological accounts of automatic processing discussed above in ch. 4.

[8] So Adrian M. Owen and Martin R. Coleman, 'Functional Neuroimaging of the Vegetative State', *Nature Reviews Neuroscience* 9 (2008), pp. 235–43 (240).

[9] B. Jennett and F. Plum, 'Persistent Vegetative State after Brain Damage', *Lancet* 1 (1972), pp. 734–37; Multi-Society Task Force on the Persistent Vegetative State, 'Medical Aspects of a Persistent Vegetative State', *New England Journal of Medicine* 330 (1994), pp. 499–508, 572–79.

[10] This was the case with the patient in Owen et al.'s 2006 study: Owen and Coleman, 'Functional Neuroimaging of the Vegetative State', p. 241.

care, because giving only physical care to a consciously aware patient would be to neglect her needs. It raised questions about possible treatment approaches: around the same time as the study by Owen et al., an early report on deep-brain (electrical) stimulation as a treatment strategy for traumatic brain injuries was published.[11] Owen et al.'s findings also had ethical and legal implications, since in various jurisdictions, permanent absence of conscious awareness is one of the criteria for permitting the withdrawal of life-sustaining treatment, including ANH.[12]

These findings proved controversial from the outset. Everyone agreed that caution and humility were needed in drawing conclusions from what were small preliminary studies, and certainly in acting on them.[13] But beyond this, some authors such as Joseph Fins welcomed them as a challenge to pay more attention to a marginalized patient population, focus more research effort on those patients and develop improved approaches to diagnosis, prognosis and care.[14] Others were more sceptical, arguing that claims of conscious awareness went beyond the evidence or were projections of hoped-for interpretations onto the experimental data.[15]

Subsequent work has answered at least some of these objections, though some critics remain sceptical that such findings can be taken as reliable evidence of conscious awareness.[16] Follow-up studies with larger patient groups have suggested that the original finding was not an isolated instance: it is claimed that at least 17–19 per cent of patients with a VS diagnosis may have the kind of covert awareness that would enable them to perform mental tasks.[17] The

[11] N. D. Schiff et al., 'Behavioural Improvements with Thalamic Stimulation after Severe Traumatic Brain Injury', *Nature* 448 (2007), pp. 600–3.

[12] Owen and Coleman, 'Functional Neuroimaging of the Vegetative State', pp. 241–42.

[13] For example, Joseph J. Fins et al., 'Neuroimaging and Disorders of Consciousness: Envisioning an Ethical Research Agenda', *American Journal of Bioethics* 8.9 (2008), pp. 3–12; Eric Racine and Emily Bell, 'Clinical and Public Translation of Neuroimaging Research in Disorders of Consciousness Challenges Current Diagnostic and Public Understanding Paradigms', *American Journal of Bioethics* 8.9 (2008), pp. 13–15.

[14] For example, Joseph J. Fins, 'Border Zones of Consciousness: Another Immigration Debate?', *American Journal of Bioethics* 7.1 (2007), pp. 51–54; Fins, 'Neuroethics and Neuroimaging'. See further Fins, *Rights Come to Mind* (New York, NY: Cambridge University Press, 2015).

[15] For example, Bernard Baertschi, 'The Burden of Self-Consciousness', *American Journal of Bioethics* 8.9 (2008), pp. 33–34; Levy, 'Going beyond the Evidence'; Kevin Chien-Chang Wu, 'Soul-Making in Neuroimaging?', *American Journal of Bioethics* 8.9 (2008), pp. 21–22.

[16] See, for example, Daniel J. Miklin and Robin N. Fiore, 'fMRI Imaging and Decision Making in Vegetative Patients: Ethics, Technology and Welfare', *AJOB Neuroscience* 6.2 (2015), pp. 49–51.

[17] Davinia Fernández-Espejo and Adrian M. Owen, 'Detecting Awareness after Severe Brain Injury', *Nature Reviews Neuroscience* 14 (2013), pp. 801–9 (806).

neuroimaging approach has been developed in various ways.[18] For example, it has been used to communicate with patients, by asking yes/no questions and instructing them to perform different mental tasks for 'yes' and 'no'. A few patients tested in this way have reportedly given verifiably correct answers and engaged in logical reasoning.[19] There has been preliminary work on enabling subjects to spell words using mental operations detectable by fMRI, though so far only with healthy volunteers.[20] It is suggested that in future, brain–computer interfaces (BCIs) might enable patients with DoCs to interact with their environment and execute intentions and commands.[21] On the strength of developments like these, Owen and others have argued that it is time for neuroimaging evidence to be used routinely in the diagnosis of DoCs and to play a part in judicial decisions about withdrawal of life-sustaining treatments.[22]

Yet important philosophical and ethical questions remain. It is still a matter of dispute what exactly neuroimaging studies are detecting, and what its ethical significance might be. The remainder of the chapter will be concerned with these debates. First, in the next section, I shall outline some philosophical and neuroscientific perspectives on consciousness and the self, and explore what bearing these might have on our understanding of DoCs and the interpretation of neuroimaging studies.

Disorders of what?

As Martha Farah has observed, beneath the technical and pragmatic questions about these studies lies a more fundamental issue, the 'little matter of consciousness': there is a risk of overlooking 'the issues of conceptual analysis and study design that make imaging consciousness different from imaging

[18] Ibid.

[19] Ibid., pp. 804–5; Martin M. Monti et al., 'Willful Modulation of Brain Activity in Disorders of Consciousness', *New England Journal of Medicine* 362.7 (2010), pp. 579–89; Adam Hampshire et al., 'Assessing Residual Reasoning Ability in Overtly Non-Communicative Patients Using fMRI', *Neuroimage: Clinical* 2 (2013), pp. 174–83.

[20] Bettina Sorger et al., 'A Real-Time fMRI-Based Spelling Device Immediately Enabling Robust Motor-Independent Communication', *Current Biology* 22 (2012), pp. 1333–38.

[21] Fernández-Espejo and Owen, 'Detecting Awareness after Severe Brain Injury', p. 808. On BCIs, see further Chapter 6.

[22] Ibid., pp. 807–8.

other brain phenomena'.[23] As with several other topics discussed in this book (such as religion, morality and free will), researchers' understandings of basic but non-self-evident concepts may shape every stage of their studies from experimental design to conclusions. This area of neuroscience, like others, is profoundly theory-laden from start to finish. This point is emphasized by Kevin Chien-Chang Wu, who seems to infer a stance of general scepticism for which he does not give particularly compelling arguments.[24] As Farah implies, however, it would seem perverse simply to write off conclusions like Owen's as wishful thinking encouraged by the human context of DoCs, if we do not apply the same level of scepticism to *behavioural* evidence of consciousness (such as a patient squeezing a clinician's hand when asked to do so).[25]

What the theory-ladenness of neuroimaging studies does mean, though, is that neuroscientific, philosophical and ethical questions are very closely entangled. If we ask what neuroimaging studies of patients with DoCs are detecting, and what its ethical significance is, both neuroscientists and philosophers will have things to say in reply.

The neuroscientist Giulio Tononi remarks: 'Everybody knows what consciousness is: it is what vanishes every night when we fall into dreamless sleep and reappears when we wake up or when we dream.'[26] This captures a commonsense understanding well, but does not specify very precisely *what* is being talked about. Philosophers of mind have gone much further in analysing the concept into various aspects; this is relevant to our topic because neuroscientific theories may not refer to all these aspects, disorders of consciousness may affect some more than others, and neuroimaging studies may assess some but not others.

This is the subject of an extensive philosophical literature, a full survey of which is well beyond the scope of this chapter.[27] In any case, when reflecting

[23] Martha J. Farah, 'That Little Matter of Consciousness', *American Journal of Bioethics* 8.9 (2008), pp. 17–19.

[24] Wu, 'Soul-Making in Neuroimaging?', citing Adina L. Roskies, 'Neuroimaging and Inferential Distance', *Neuroethics* 1.1 (2008), pp. 19–30.

[25] Farah, 'That Little Matter of Consciousness', pp. 18–19.

[26] Giulio Tononi, 'Consciousness as Integrated Information: A Provisional Manifesto', *Biological Bulletin* 215 (2008), pp. 216–42 (216). He continues, 'It is also all we are and all we have: lose consciousness and, as far as you are concerned, your own self and the entire world dissolve into nothingness' (ibid.).

[27] For a helpful overview, see Robert van Gulick, 'Consciousness', in Edward N. Zalta (ed.), *The Stanford Encyclopedia of Philosophy* (Spring 2014 Edition), online at http://plato.stanford.edu/archives/spr2014/entries/consciousness/ (accessed 9 May 2016).

on neuroimaging studies and their ethical significance, it would complicate matters unnecessarily to use too fine-grained a taxonomy. However, one key distinction has proved important in the ethical discussion of DoCs. This is Ned Block's distinction between *phenomenal consciousness* (or P-consciousness) and *access consciousness* (or A-consciousness).[28] P-consciousness is about experience: the qualitative character of a state, or what it is like to be in that state. A-consciousness has to do with the availability of mental representations: 'A representation is A-conscious if it is broadcast for free use in reasoning and for direct "rational" control of action (including reporting).'[29] Block maintains that P- and A-consciousness are conceptually distinct and it is conceivable to have either one without the other. He also distinguishes P-consciousness from two cognitive states: *self-consciousness*, or 'the possession of the concept of the self and the ability to use this concept in thinking about oneself',[30] and what he calls *monitoring consciousness* – that is, a meta-mental condition of being aware of one's own state.[31] Block resists the identification of P-consciousness with either self- or monitoring consciousness, because he maintains that it is distinct from cognition.[32]

Like philosophical theories, neuroscientific accounts of consciousness are legion. These include attempts to identify the neural correlates of consciousness,[33] proposals about the necessary or sufficient neural conditions for it to exist,[34] explanations of how brain activity can give rise to it,[35] and claims about its function – why it exists.[36] One of the most prominent is the *global workspace theory*.[37]

[28] Ned Block, 'On a Confusion about a Function of Consciousness', *Behavioral and Brain Sciences* 18.2 (1995), pp. 227–47; reprinted in revised form as 'Concepts of Consciousness' in Ned Block, *Consciousness, Function and Representation* (Cambridge, MA: MIT Press, 2007), pp. 275–96. Subsequent citations will be to the 2007 version.

[29] Block, *Consciousness, Function and Representation*, p. 278.

[30] Ibid., p. 287.

[31] Ibid., pp. 288–89; the term 'meta-mental' is from van Gulick, 'Consciousness', section 2.2.

[32] Block, *Consciousness, Function and Representation*, p. 276.

[33] For example, Francis Crick and Christof Koch, 'A Framework for Consciousness', *Nature Neuroscience* 6.2 (2003), pp. 119–26.

[34] For example, Steven Laureys, 'The Neural Correlate of (Un)awareness: Lessons from the Vegetative State', *Trends in Cognitive Sciences* 9.12 (2005), pp. 556–59.

[35] For example, Bernard J. Baars, 'Global Workspace Theory of Consciousness: Toward a Cognitive Neuroscience of Human Experience', *Progress in Brain Research* 150 (2005), pp. 45–53; Tononi, 'Consciousness as Integrated Information'.

[36] For example, Baars, 'Global Workspace Theory'.

[37] Ibid.; Stanislas Dehaene et al., 'Conscious, Preconscious, and Subliminal Processing: A Testable Taxonomy', *Trends in Cognitive Sciences* 10.5 (2006), pp. 204–11.

In this theory, information (for example, sensory content) becomes conscious as it is made accessible to a widely distributed range of 'expert' systems or networks in the brain with different specialized processing functions. Since these networks have limited processing capacity, content has to 'compete' for access to them, and not all of it becomes consciously accessible.[38] Bernard J. Baars likens the global workspace of the brain to a theatre, in which consciousness can be compared to 'a bright spot on the stage of immediate memory, directed there by a spotlight of attention under executive guidance'.[39] Global workspace is a theory about how the brain gives rise to consciousness, but also about its function: to give many specialized networks in the brain access to the same information and to integrate their functioning as they process and use that information.[40]

Stanislas Dehaene and his colleagues argue that conscious access to information requires not only wakefulness (or 'vigilance') and bottom-up activation of relevant brain areas (such as primary sensory cortices), but also the extension of this activation to higher associative cortices, particularly in the parietal and frontal lobes. The involvement of these higher associative areas gives rise to 'reverberating' activation of long-distance neural networks, which can be maintained as long as desired, and is available to guide diverse actions including verbal reporting.[41]

This predicts that A-consciousness should be correlated with activation of higher associative cortices, and there is some empirical evidence for this.[42] It also seems consistent with neuroimaging findings that higher associative cortices are active in MCS, whereas in VS, primary sensory activity is 'functionally disconnected' from higher associative areas.[43] Lack of evidence for this higher associative activity was one reason for Neil Levy's initial doubtfulness about whether Owen et al. really had detected consciousness in the patient described in their initial report.[44] In later publications, however, Owen and

[38] This can be seen in phenomena like inattentional blindness, in which subjects paying attention to one visual input fail to observe another one. For a vivid example, see Daniel J. Simons and Christopher F. Chabris, 'Gorillas in our Midst: Sustained Inattentional Blindness for Dynamic Events', *Perception* 28.9 (1999), pp. 1059–74.

[39] Baars, 'Global Workspace Theory', p. 46.

[40] Ibid., p. 47.

[41] Dehaene et al., 'Conscious, Preconscious, and Subliminal Processing'.

[42] Ibid., p. 205.

[43] Laureys et al., 'Cortical Processing of Noxious Somatosensory Stimuli'; Schiff et al., 'fMRI Reveals Large-Scale Network Activation'.

[44] Levy, 'Going beyond the Evidence', p. 20.

his colleagues have argued that the mental tasks performed by several patients demonstrated diverse forms of top-down processing.[45] Moreover, their study of logical reasoning in a DoC patient showed widespread right hemisphere activity correlated with the set task; they judge it 'likely ... that the patient engaged in the reasoning task, bringing those frontal-lobe resources that were still available on-line in response to more complex problems'.[46] These results go some way towards meeting the expectations that the global workspace theory suggests.

Global workspace theory deals most obviously with A-consciousness, and indeed Dehaene et al. more or less restrict their account to this aspect.[47] Baars, however, also offers an account of first-person experiential aspects.[48] He attributes the first-person perspective to what he calls 'contexts': unconscious information processing that maps the body and its relation to the external world.[49] He also proposes a role for 'self-systems' in generating subjective consciousness, drawing on Michael Gazzaniga's proposal that there is a 'narrative interpreter' of actions and events, located in the left hemisphere.[50] The role of narrative interpretation in consciousness has also been argued by others such as Daniel Dennett,[51] and features prominently in Daniel Wegner's argument for the illusory nature of conscious will, discussed in Chapter 4.[52]

[45] For example, Fernández-Espejo and Owen, 'Detecting Awareness after Severe Brain Injury', p. 805. In this connection they refer to the patient's *cognitive* status and competency, and as the philosophical and neuroscientific theories emphasize, cognition and consciousness do not necessarily go together. However, the range of functions they describe, including attention, memory, expression of preferences and accessing 'information about [one's] bodily state', appear highly suggestive of elements of both A- and P-consciousness.

[46] Hampshire et al., 'Assessing Residual Reasoning Ability', p. 182. The specific areas involved in the reasoning task in healthy volunteers had been damaged by the patient's injury, hence the suggestion that in their place he was recruiting surviving brain areas for the task.

[47] So Melanie Boly, Marcello Massimini and Giulio Tononi, 'Theoretical Approaches to the Diagnosis of Altered States of Consciousness', *Progress in Brain Research* 177 (2009), pp. 383–98 (385); see Dehaene et al., 'Conscious, Preconscious, and Subliminal Processing'.

[48] What exactly he means by this is not entirely clear from his account, but it seems to include at least some of what Block means by P-consciousness, as well as elements of the cognitive aspects that Block differentiates from P-consciousness.

[49] Baars, 'Global Workspace Theory', p. 49.

[50] Ibid., p. 50; see, for example, Matthew Roser and Michael S. Gazzaniga, 'Automatic Brains – Interpretive Minds', *Current Directions in Psychological Science* 13.2 (2004), pp. 56–59. What Baars seems to mean by 'subjective consciousness' is what van Gulick ('Consciousness', section 2.2) calls a meta-mental state, or a combination of what Block (*Consciousness, Function and Representation*, pp. 287–89) describes as self-consciousness and monitoring consciousness.

[51] For example, Daniel Dennett, *Consciousness Explained* (Boston, MA: Little, Brown and Co., 1991); see van Gulick, 'Consciousness', section 9.4.

[52] Daniel M. Wegner, *The Illusion of Conscious Will* (Cambridge, MA: MIT Press, 2002).

Global workspace is far from the only theoretical account available. Phenomenal consciousness is accounted for differently, for example, by Guilio Tononi's information-theoretical model, in which consciousness simply is 'integrated information'. By this, Tononi means that the quantity of consciousness in a system is the amount of information it generates, over and above what could be generated by its component parts.[53] The *quality* of a conscious experience (roughly what Block means by P-consciousness) is a function of the informational relationships between the elements of the system.[54] Tononi and his colleagues argue that this theory is more successful than others at explaining neuroscientific data, including evidence from some situations of loss of consciousness.[55]

Differences between these rival theories seem a long way from being resolved. Until such time as they are, theoretical perspectives might generate illuminating insights and questions about experimental findings (as in Levy's critique of Owen et al., outlined above), but these insights should be used with caution. It would be dangerously premature, for example, to conclude that DoC patients *cannot* be truly conscious just because their brain activity does not conform to predictions generated by one or other of the theories.

We are accustomed to thinking of ourselves as conscious *selves*, yet neuroscientific theories of consciousness often sideline or even bracket out discussion of the self.[56] One neuroscientist who does not is Antonio Damasio. Indeed, he approaches consciousness by way of the self, proposing that 'conscious minds arise when a self process is added onto a basic mind process'.[57]

Damasio understands the self not as a thing but a process generated by the brain. It has ancient origins in evolutionary history, and its original function was to do with homeostasis: the maintenance of the body's internal environment within the limits necessary to sustain life and health. His

[53] Tononi, 'Consciousness as Integrated Information', pp. 218–19.

[54] Tononi represents this mathematically as a many-dimensional 'qualia space' in which each quale or other experience can be represented as a shape defined by the relationships between the elements contributing to it: ibid., pp. 224–32.

[55] Boly et al., 'Theoretical Approaches'.

[56] For example, Crick and Koch, 'A Framework for Consciousness'; though as noted above, Baars is at least a partial exception: 'Global Workspace Theory', p. 50.

[57] Antonio Damasio, *Self Comes to Mind: Constructing the Conscious Brain* (London: Vintage, 2012), p. 8.

model is threefold. First (and most evolutionarily ancient) is the 'protoself', concerned with mapping the body, its internal states and the external world, and generating 'primordial feelings', such as pain and pleasure, associated with that mapping information. Next comes the 'core self', concerned with the relationships between the self and other objects, and with action in response to those objects. Finally, there is the 'autobiographical self' (the most recent in evolutionary history), which involves a wide range of systems and circuits in the brain. Some degree of consciousness emerges with the core self, but the full richness of conscious life requires an autobiographical self.

Damasio believes many animals have core selves, and some mammalian species besides humans have both core and autobiographical selves.[58] When it comes to patients with DoCs, however, his model leads him to take a sceptical view about claims of conscious awareness: 'in spite of the significance of [the findings reported by Owen et al.] … I am reluctant to regard them as evidence for conscious communication'.[59] Ned Block attributes Damasio's reluctance to his insistence (mistaken in Block's view) that P-consciousness depends on 'inflated self-consciousness', and is unimpressed: '[Damasio] owes us an explanation of why he thinks chickens are conscious even though … question-answering patients are not'.[60] In any event, the more recent reports by Owen and others should arguably overcome Damasio's scepticism. A patient who can use mental imagery to answer questions correctly about his own and others' names, his location and the year seems to be demonstrating an awareness of his own identity, his relationship to others and his temporal and spatial context;[61] on either view, this should surely count as evidence of *conscious* awareness.

So far, so philosophical and neuroscientific. However, throughout this book my aim is to explore the difference it makes to engage theologically with neuroethical questions of diverse kinds. Therefore, the next section will briefly explore one way of gaining a Christian theological perspective on these debates about consciousness and the self.

[58] Ibid., p. 26.

[59] Ibid., p. 162.

[60] Ned Block, 'What Was I Thinking?', *New York Times*, 28 November 2010. For further critique of Damasio's account, see Lisa Bortolotti, 'Review of *Self Comes to Mind* by Antonio Damasio', *Cognitive Neuropsychiatry* 17.2 (2012), pp. 191–95.

[61] Fernández-Espejo and Owen, 'Detecting Awareness after Severe Brain Injury', p. 805.

Selves, souls and consciousness: Theological perspectives

If one asks how Christian theology might engage with these theories of consciousness and the self, it seems intuitively obvious that the answer will have something to do with a Christian concept of the soul. And indeed, talk of the soul can often be found in these discussions. Neuroscientists who theorize about consciousness sometimes imagine that their work discredits 'religious' views of the soul. A familiar example is Francis Crick's 'Astonishing Hypothesis ... that "you," your joys and your sorrows, your memories and your ambitions, your sense of personal identity and free will, are in fact no more than the behavior of a vast assembly of nerve cells and their associated molecules.'[62] Crick takes this view to indicate 'the true nature of the human soul', and to be 'in head-on contradiction to the religious beliefs of billions of human beings alive today'.[63]

By and large, the 'religious' view of the soul targeted by such critics is some kind of substance dualism. Theologians have not been slow to point out that those who take aim at this target are attacking what Ted Peters calls a straw theology.[64] Such a response may be made from biblical, philosophical, systematic theological and other perspectives. For example, Joel Green has argued in many publications that body-soul dualism has relatively little support from biblical texts in either Testament. While there is no unified 'biblical' view of the person, nor for that matter a holistic 'Hebrew' view and a dualistic 'Greek' one, Green maintains that the anthropology of New Testament authors such as Luke and Paul is 'fundamentally social and wholistic

62 Francis Crick, *The Astonishing Hypothesis: The Scientific Search for the Soul* (New York: Charles Scribner's Sons, 1994), p. 3.

63 Ibid., pp. 4, 261. It would of course take much more than a proposal about the neural *correlates* of consciousness, which was Crick's own contribution to the field, to substantiate this 'astonishing hypothesis'.

64 Ted Peters, 'Resurrection of the Very Embodied Soul?', in Robert John Russell et al. (eds.), *Neuroscience and the Person: Scientific Perspectives on Divine Action* (Vatican City: Vatican Observatory Publications/Berkeley, CA: Center for Theology and the Natural Sciences, 1999), pp. 305–26. This disavowal of substance dualism is a dominant theme of numerous edited collections on theology and neuroscience: for example, Warren S. Brown, Nancey Murphy and H. Newton Moloney (eds.), *Whatever Happened to the Soul? Scientific and Theological Portraits of Human Nature* (Minneapolis, MN: Fortress, 1998); Joel B. Green (ed.), *What about the Soul? Neuroscience and Christian Anthropology* (Nashville, TN: Abingdon, 2004); Malcolm Jeeves (ed.), *From Cells to Souls – and Beyond: Changing Portraits of Human Nature* (Grand Rapids, MI: Eerdmans, 2004); Russell et al., *Neuroscience and the Person*.

in its presentation.'[65] From a philosophical perspective, Nancey Murphy is well known for advocating a non-reductive physicalist philosophy of mind, deploying key concepts, including supervenience, emergence and top-down causation, as well as drawing on the biblical scholarship of Green and others to argue for the consonance of this non-reductive physicalist view with a biblically grounded theological anthropology.[66] From systematic theology, Ted Peters (for example) argues against dualist eschatologies, drawing on Pannenberg and to a lesser extent Barth to argue that Christian eschatological hope is better represented as 'God's resurrection of the whole person.'[67]

Few theologians who engage with neuroscience, in short, would wish to defend a substance-dualist account of the soul.[68] Within this broad anti-dualist consensus, there is of course a range of views about how the human person *should* be understood. Nancey Murphy, well known for her non-reductive physicalist position, thinks it is possible for Christians to say that '[n]one of us has a soul and we all get along perfectly well.'[69] Alan Torrance, while rejecting substance dualism, is also highly critical of the physicalist views of Murphy and others. Instead, he advocates the 'pluralist' view proposed by the philosopher of science Nancy Cartwright, in which a variety of higher-order patterns with genuine causal efficacy emerge from physical processes; the mental life of human persons could be one among many such emergent phenomena in what Cartwright calls a 'dappled world.'[70] (It is not clear how distant this position really is from the view of emergence and top-down causation proposed by Murphy, though Torrance is certainly critical of Murphy's positions on matters such as supervenience and mental causation, which were discussed above in Chapter 4.) In sharper contrast with Murphy, Keith Ward – drawing on

65 Joel B. Green, 'Restoring the Human Person: New Testament Voices for a Wholistic and Social Anthropology', in Russell et al. (eds.), *Neuroscience and the Person*, pp. 3–22 (quotation at 3).

66 See, for example, Nancey Murphy, *Bodies and Souls, or Spirited Bodies* (Cambridge: Cambridge University Press, 2006), and for a recent overview of her position, Murphy, 'Do Humans Have Souls? Perspectives from Philosophy, Science, and Religion', *Interpretation* 67.1 (2013), pp. 30–41. Murphy's account of supervenience, emergence and top-down causation, and their relevance for free will in particular, were discussed above in ch. 4.

67 Peters, 'Resurrection of the Very Embodied Soul?', p. 322.

68 There are a few exceptions, such as Stewart Goetz, 'Substance Dualism', in Joel B. Green and Stuart L. Palmer (eds.), *In Search of the Soul: Four Views of the Mind-Body Problem* (Downers Grove, IL: InterVarsity Press, 2005), pp. 33–60.

69 Murphy, 'Do Humans Have Souls?', p. 30.

70 Alan Torrance, 'What Is a Person?', in Jeeves, *From Cells to Souls*, pp. 199–222, citing Nancy Cartwright, *The Dappled World: A Study of the Boundaries of Science* (Cambridge: Cambridge University Press, 1999).

Thomas Aquinas among others – thinks that a concept of the soul does do real and necessary theological work for a Christian anthropology. Under Aquinas' influence, he is also more ready than some to think of the soul continuing to exist, in God's care, after the death of the body.[71] Yet his view of the soul is robustly non-dualist: '[b]elievers should not wish to deny that the soul *is* the embodied person we think and see and know in the world'.[72]

One unfortunate aspect of these discussions is that the theologians' chief target often seems to be the same 'straw theology' attacked by their materialist opponents, namely, some kind of dualism.[73] Of course there are some good reasons for this: dualism is after all popularly supposed to be the 'religious' position, which materialists like Crick think they have discredited, so it is no doubt important to emphasize that their attack is irrelevant to Christian theology. Also, there may still be an unhealthy amount of dualism around in Christian theology, as Fergus Kerr argued some years ago[74] – in which case, it is certainly worth emphasizing its problems as a Christian view of the human person. But if the discussion becomes too focused on opposing dualism and putting in its place some kind of monist or physicalist account such as Murphy's, other resources from the Christian tradition that could enrich the discussion can end up somewhat sidelined.[75]

That caveat aside, there are good reasons to think that in theological anthropology's engagement with neuroscience, the language of body and soul will be best understood – as Murphy puts it – in an 'aspective' rather than a 'partitive' way.[76] That is, 'body' and 'soul' denote the whole human person, seen

[71] Keith Ward, *In Defence of the Soul* (Oxford: Oneworld, 1998), pp. 147–52.

[72] Ibid., p. 151, emphasis original.

[73] See, for example, D. Gareth Jones, 'The Emergence of Persons', and Diogenes Allen, 'Persons in Philosophical and Biblical Perspectives', in Jeeves, *From Cells to Souls*, pp. 11–33 and 165–78 respectively; also Green refers to 'the central place body-soul dualism has had in the Christian tradition from the patristic period on': 'Restoring the Human Person', p. 22.

[74] Fergus Kerr, 'The Modern Philosophy of Self in Recent Theology', in Russell et al. (eds.), *Neuroscience and the Person*, pp. 23–40.

[75] One under-used resource, perhaps, is Aristotelian-Thomist hylomorphism, which for a good part of Western Christian history was *the* mainstream view. It is not of course absent altogether from these discussions: as we have seen, Keith Ward's account draws significantly on Aquinas, and not long before his death Stephen Happel outlined a highly creative attempt to read a Thomist account of the soul together with Husserlian phenomenology and neuroscience: Stephen Happel, 'The Soul and Neuroscience: The Empirical Conditions for Human Agency and Divine Action', in Russell et al. (eds.), *Neuroscience and the Person*, pp. 281–304.

[76] Murphy, 'Do Humans Have Souls?', p. 33, citing James D. G. Dunn, *The Theology of Paul the Apostle* (Grand Rapids, MI: Eerdmans, 1998), p. 54.

from different perspectives. In the language of Barth's account, they should be 'understood, not as two parts, but as two moments of the indivisibly one human nature, the soul as that which quickens and the body as that which is quickened and lives. It is to this concrete monism that we found ourselves guided by the biblical view and the biblical concept of the "soul"'.[77]

Such a view will be hospitable to insights from neuroscientific theories of consciousness and the self, such as those surveyed in the last section, to the extent that those insights are scientifically robust and do not include covert materialist presuppositions. For example, should Damasio's threefold model of the self-stand the tests of time and empirical investigation, this theological view ought to have no particular difficulty understanding such a human self as a *creaturely* self, made, known, loved and called to an eternal hope by God. Again, should a global workspace theory of consciousness prove scientifically robust, then consciousness in that sense will be understood theologically as one way in which a human creature who is 'soul of [her] body' (Barth) functions in relation to God, others, herself and her environment.

It will not be for theology to judge the scientific merits of rival theories, but theologians may certainly draw attention to covert or overt (a-)theological claims made by neuroscientists who overstep the bounds of their scientific competence – as in the quotation from Crick at the beginning of this section. Theologians might also have a good vantage point from which to raise critical questions about the various theories. For example, Robert van Gulick observes that Tononi's integrated information theory of consciousness is substrate-neutral,[78] in which case it might lend itself to transhumanist speculations about the uploading of human minds to computers.[79] Theologians with an anti-dualist view of body and soul may discern a highly questionable kind of dualism *redivivus* in these speculations.

Theological accounts of the kind surveyed in this section have often emphasized the social and relational aspects of human personhood: to be a

[77] Karl Barth, *Church Dogmatics* (G. W. Bromiley and T. F. Torrance (eds.), 13 vols., Edinburgh: T&T Clark, 1956–75), vol. 3.2, p. 393. Moreover, Barth emphasizes that our unity as human creatures depends on the creating and sustaining work of God: 'it is the Spirit, i.e., the immediate action of God Himself, which grounds, constitutes and maintains man as soul of his body. It is thus the Spirit that unifies him and holds him together as soul and body' (ibid.). See further Ray S. Anderson, 'On Being Human: The Spiritual Saga of a Creaturely Soul', in Brown et al. (eds.), *Whatever Happened to the Soul?*, pp. 175–94.

[78] Van Gulick, 'Consciousness', sect. 9.6.

[79] Transhumanism, including mind uploading, is briefly discussed in Chapter 6, below.

human creature, 'soul of one's body', is to be in relationship with God and one's fellow-creatures.[80] In itself, this is a welcome emphasis, a valuable corrective to the individualism which was perhaps another legacy of the Cartesian ego.[81] But some authors express this in terms of the human *capacity* for relationships, and this has its problems.

For example, Warren Brown writes of human 'soulishness' as an emergent property whose 'critical feature ... is the capacity for deep and rich experiences of *personal relatedness*':[82] relatedness to others, and '[t]he capacity for, and experiences of, *relatedness to God*'.[83] Now, it is well known that if personhood is thought to depend on human capacities (such as the capacity for relatedness), this brings into question the moral status of humans who are severely impaired in those capacities, as we shall see in the following sections.[84] But by describing the 'capacity for ... *relatedness to God*' as an emergent 'soulish' property of human brains, Brown also seems to make humans' *theological* status dependent on our capacities. His view would seem to bring into question the standing before God of humans whose capacity for relatedness was severely impaired, a troubling conclusion for Christians who wish to affirm that nothing in creation can separate us from God's love in Christ (cf. Rom. 8.38-39). Brown recognizes this problem, and attempts a solution by suggesting that a community or a more 'relationally competent' person may evoke a response even from one whose capacity for relatedness is impaired. While this is a helpful and important point, it is only a partial solution, because the standing of those whose impairments leave little or no capacity for evoked responses remains in doubt.[85] Christians who wish to resist such troubling conclusions may find that

[80] For example, Green, 'Restoring the Human Person'; Torrance, 'What Is a Person', drawing on John Macmurray, John Zizioulas and especially Karl Barth. It should be noted that in some of the literature, connections are made between human personhood as relational being and 'social Trinitarianism' in its various forms, a set of connections that has been vigorously disputed in recent years: see, for example, Karen Kilby, 'Perichoresis and Projection: Problems with Social Doctrines of the Trinity', *New Blackfriars* 81 (2000), pp. 432–45.

[81] So Kerr, 'The Modern Philosophy of Self'.

[82] Warren S. Brown, 'Neurobiological Embodiment of Spirituality and Soul', in Jeeves, *From Cells to Souls*, pp. 58–76 (67, emphasis original).

[83] Ibid., p. 68, emphasis original.

[84] Gareth Jones is one Christian author who follows this logic through. Though he does not state this conclusion firmly, he apparently leans towards the view that PVS patients have died as persons and it is therefore permissible to withdraw ANH: 'The Emergence of Persons', pp. 16–20, 30–31.

[85] For an extended reflection on the theological questions raised by encounters with such profound impairment, see Hans S. Reinders, *Receiving the Gift of Friendship: Profound Disability, Theological Anthropology, and Ethics* (Grand Rapids, MI: Eerdmans, 2008).

an account more like that of Barth, who emphasizes that it is from first to last *God's* action which constitutes the human person as 'soul of [her] body', gives them more resources to do so.

The implications of functional neuroimaging: Ethical debates

The previous two sections have surveyed philosophical, neuroscientific and theological perspectives on some of the theoretical questions raised by disorders of consciousness and functional neuroimaging studies. The philosophical and neuroscientific discussion introduced concepts and distinctions that play an important part in some ethical analyses of the neuroimaging studies. The theological discussion explored briefly how a Christian anthropology might relate to philosophical and scientific perspectives on humans as conscious selves. The discussion of Christian anthropology, and particularly the language of the soul, in relation to neuroscience in the last section began to raise ethical questions. In particular, I suggested that some ways of thinking about the soul might prove to have problematic consequences for Christian ethical thinking about patients with DoCs. In the remainder of the chapter, the focus shifts back to the ethical questions themselves. In this section, I offer a critical survey of some philosophical accounts – particularly focusing on analyses by Julian Savulescu and his colleagues – while in the remaining two sections I shall attempt to reframe the ethical discussion theologically.

In ethical debates about DoCs, it is widely believed that consciousness marks a crucial moral boundary, giving a strong reason to sustain the lives of those who are conscious or have the capacity for consciousness.[86] Fins et al. hint at one line of reasoning that frequently underpins this view, when they claim that the cost of failing to recognize consciousness in patients with DoCs is 'neglected personhood'.[87] If personhood is what entitles an individual to the kind of moral regard we typically owe other human beings and consciousness is a sufficient condition for personhood,[88] then robust evidence of consciousness

[86] As Guy Kahane and Julian Savulescu observe: 'Brain Damage and the Moral Significance of Consciousness', *Journal of Medicine and Philosophy* 34 (2009), pp. 6–26 (9).

[87] Fins et al., 'Neuroimaging and Disorders of Consciousness', p. 4.

[88] This seems to be implicit in what Fins et al. say; others, of course, might maintain that it is a necessary but not sufficient condition.

is a good reason to attribute personhood to an individual.[89] Moreover, as Fins et al. suggest, if we fail to notice the evidence, we may 'neglect' her personhood. For Fins, recognizing the 'neglected personhood' of patients with DoCs would have important consequences for their care, since many of these patients do not currently receive the kind of care that befits persons.[90] He and his colleagues also argue that justice requires this patient population to be enfranchised, as they put it, in scientific and clinical research, a claim that has met with some critical responses.[91] Finally (and most ethically charged), it would have an important bearing on decisions about the withdrawal of ANH.[92]

Julian Savulescu and others have challenged this line of thought in various ways.[93] They have asked in what sense neuroimaging studies give evidence of consciousness; they have questioned whether that evidence means these patients are *persons*, and they have challenged common assumptions about the ethical implications even if they are.

As we have seen, Levy was initially sceptical about the early neuroimaging findings.[94] However, he and Savulescu later conceded (their word) that Owen's 2006 study offered 'impressive' evidence of consciousness.[95] As to what kind of consciousness this might be, Savulescu and his colleagues draw on Block's distinctions between access, phenomenal and self-consciousness, and suggest that these different forms could come apart in DoCs. Levy and Savulescu conclude that the patient reported in Owen's 2006 study, who could

[89] This basic line of thinking is extremely common in bioethics. Its origins as an approach to bioethical reasoning specifically can be traced to authors such as Mary Anne Warren, 'On the Moral and Legal Status of Abortion', *The Monist* 57.1 (1973), pp. 43–61, and Joseph F. Fletcher, 'Four Indicators of Humanhood: The Enquiry Matures', *Hastings Center Report* 4.6 (1974), p. 4; though as we shall see, it draws on a much longer intellectual history than this.

[90] Fins, 'Border Zones of Consciousness', p. 52.

[91] Fins et al., 'Neuroimaging and Disorders of Consciousness', pp. 8–9; for critical responses, see, for example, Stephanie J. Bird, 'Ethical Challenges in Research: Another Look', *American Journal of Bioethics* 8.9 (2008), pp. 15–17; Rob Schwartz and Mirra Schwartz, 'The Risks of Reducing Consciousness to Neuroimaging', *American Journal of Bioethics* 8.9 (2008), pp. 25–26.

[92] Though Fins distances himself from a pro-life position: 'Border Zones of Consciousness', p. 53.

[93] Kahane and Savulescu, 'Brain Damage and the Moral Significance of Consciousness'; Levy, 'Going beyond the Evidence'; Neil Levy and Julian Savulescu, 'Moral Significance of Phenomenal Consciousness', *Progress in Brain Research* 177 (2009), pp. 361–70; Dominic Wilkinson, Guy Kahane and Julian Savulescu, '"Neglected Personhood" and Neglected Questions: Remarks on the Moral Significance of Consciousness', *American Journal of Bioethics* 8.9 (2008), pp. 31–33; D. J. Wilkinson et al., 'Functional Neuroimaging and Withdrawal of Life-Sustaining Treatment from Vegetative Patients', *Journal of Medical Ethics* 35 (2009), pp. 508–11.

[94] Levy, 'Going beyond the Evidence'.

[95] Levy and Savulescu, 'Moral Significance of Phenomenal Consciousness', p. 366.

perform mental imagery tasks when instructed, was showing evidence of A-consciousness, and probably had some level of P-consciousness also.[96]

They deny that P-consciousness plus basic A-consciousness suffices for personhood. To be a person, one must also be capable of 'extended and self-referential mental states', plans, projects and future-oriented preferences. This requires a sophisticated kind of temporally extended A-consciousness, and self-consciousness.[97] Levy and Savulescu doubt whether Owen's 2006 report demonstrated all these capacities, but allow that neuroimaging evidence that did so would 'put the moral status of PVS patients beyond any doubt'.[98] As we saw earlier, Owen and others maintain that more recent imaging studies have shown that some patients can answer questions correctly, indicate that they are not in pain and recall some events that have happened since their injuries. Provided this evidence is robust and correctly interpreted, it would certainly appear to meet Levy and Savulescu's stipulations. If so, then it seems that some patients with DoCs are, by their criteria, *persons*.

What might the ethical implications of this be? Here again, Savulescu and his colleagues challenge common assumptions. According to Levy and Savulescu, although possession of P-consciousness is insufficient for personhood, it does make a being a 'moral patient' – that is, a being with *interests*, whose welfare matters.[99] Kahane and Savulescu also claim that the moral significance of consciousness consists in the possession of interests.[100]

They then parse the ethical implications of four possible scenarios. First, patients with DoCs might lack P-consciousness, A-consciousness and self-consciousness (the authors use the term 'sapience' as shorthand for the latter two). Such patients would be in a truly vegetative state, with no conscious self or experience, and no interests. Therefore, it could not be against their (present) interests to end their lives, for example, by withdrawing ANH. Second, they might lack P-consciousness but be sapient. This empirically

[96] Ibid.
[97] Ibid., p. 368.
[98] Ibid., p. 369.
[99] Ibid., p. 366.
[100] Kahane and Savulescu, 'Brain Damage and the Moral Significance of Consciousness', p. 11. By interests they mean 'roughly … what affect a being's good, welfare, or well-being' (ibid., p. 23 n. 7). They identify different kinds: *experiential*, including interests in experiencing pleasures and avoiding pain; *desiderative*, or interests in having one's desires and preferences satisfied; *objective*, concerned with objective goods of being this kind of being.

unlikely but conceivable scenario is morally ambiguous and perplexing to consider: if P-consciousness is what underwrites interests, such patients may lack interests, but might qualify as *persons* whose preferences should be respected even in the absence of interests.[101] Third, they might be P-conscious but not sapient, the condition of patients in MCS. Such patients may have interests (including not being in pain), but are not self-aware, have little psychological continuity over time and lack the desire to continue living. 'It is thus not clear', Kahane and Savulescu conclude, 'that [such a patient] would be significantly harmed if her life ended earlier than was possible'.[102] Levy and Savulescu hold that such patients would not qualify as persons and therefore would lack a right to life.[103] Fourth, patients might be both P-conscious and sapient, a condition best understood as a kind of fully locked-in state, without even the minimal motor ability to communicate by blinking or moving an eye. Kahane and Savulescu assert that such a condition would be worse than death. They acknowledge that such patients would have a view about this; the problem is that they would have no way to express it. So until imaging or other technologies make it possible to ask such patients about their own preferences, Kahane and Savulescu argue that there may be a moral *obligation* to end their lives, on the assumption that their lives are so awful that death is preferable.[104]

According to Kahane and Savulescu, past as well as present interests may be relevant. If patients who now lack P-consciousness had the desire, when they were healthy, not to continue living should they enter PVS or MCS, then that past interest might give grounds for ending their lives now that they are in such a state. How would we know they had such a desire? Most obviously, if they expressed it in an advance directive. But Kahane and Savulescu also seem to suggest that we should assume it, even in the absence of an advance directive, because many people say in surveys that they would not want to go on living in such a state.[105] Finally, resource allocation arguments are used to advocate withdrawal of life-sustaining treatments from patients with DoCs on the grounds that those treatment resources might be better used elsewhere.[106]

[101] Ibid., p. 17.

[102] Ibid., p. 18.

[103] Levy and Savulescu, 'Moral Significance of Phenomenal Consciousness', pp. 367–68.

[104] Kahane and Savulescu, 'Brain Damage and the Moral Significance of Consciousness', p. 21.

[105] Ibid., p. 15.

[106] For example, Wilkinson et al., 'Functional Neuroimaging and Withdrawal of Life-Sustaining Treatment'.

In short, Savulescu and his colleagues state several reasons for ending the lives of patients with DoCs even if they have some measure of consciousness, and argue that higher levels of consciousness may generate an obligation to do so.

Some critical remarks about all this are in order. First, Savulescu and his various co-authors frame the discussion almost exclusively in terms of interests, dismissing the possibility that there might be other reasons why human lives and/or consciousness matter morally. Kahane and Savulescu do briefly mention one such reason, which they rather inaccurately label 'sanctity of life', citing a short Catholic magisterial document on the withdrawal of ANH.[107] At the outset of their discussion, Kahane and Savulescu bracket out this position on the grounds that disputes about the treatment of patients with DoCs typically turn on whether or not they are conscious, so these disputes cannot be about 'sanctity of life'.[108] But that is too quick. It is quite possible that many of those appealing to evidence for consciousness *also* hold that the human dignity of patients with DoCs gives a powerful moral reason for sustaining their lives.[109] If so, consciousness may still be morally significant because respecting a patient's human dignity may require different actions depending on whether she is conscious. But permanent loss of consciousness will not provide sufficient grounds to conclude that it is permissible to end her life. Kahane and Savulescu do not argue against this position, but simply sidestep it.

Second, Savulescu and his colleagues make some contentious claims about *what* is in the interests of patients with DoCs, and about the kinds of life that are not worth living. For example, Kahane and Savulescu claim that a patient

[107] That document does not use the phrase 'sanctity of life', but instead describes PVS patients as 'person[s] with fundamental human dignity': Congregation for the Doctrine of the Faith, 'Responses to Certain Questions of the United States Conference of Catholic Bishops Concerning Artificial Nutrition and Hydration' (1 August 2007), online at http://www.vatican.va/roman_curia/congregations/cfaith/documents/rc_con_cfaith_doc_20070801_risposte-usa_en.html (accessed 14 May 2016). This was written as a clarification of John Paul II, 'Address of John Paul II to the Participants in the International Congress on "Life-Sustaining Treatments and Vegetative State: Scientific Advances and Ethical Dilemmas"' (20 March 2004), para. 4, online at http://w2.vatican.va/content/john-paul-ii/en/speeches/2004/march/documents/hf_jp-ii_spe_20040320_congress-fiamc.html (accessed 4 November 2016). These documents have prompted considerable discussion and controversy among Catholic moral theologians, particularly in the United States: for helpful surveys and commentary, see Jeffrey P. Bishop and Elliott Louis Bedford (eds.), Theme Issue, 'Medically Assisted Nutrition and Hydration: The Vegetative State and Beyond', *Christian Bioethics* 17.2 (2011).

[108] Kahane and Savulescu, 'Brain Damage and the Moral Significance of Consciousness', p. 9.

[109] Perhaps, for example, they focus on consciousness in an attempt to persuade opponents who would not accept their claims about human dignity, or because they are arguing in a legal context which more or less obliges them to treat consciousness as the key criterion.

who is both P-conscious and sapient, '[b]eing aware of his condition and with little hope of communication with others, ... cannot pursue most of his desires and personal projects and is cut off from most of the objective goods that make for a meaningful human life'.[110] This is backed up with an emotively powerful testimony by a partially locked-in patient about the awfulness of his condition. They acknowledge that surveys of locked-in patients report quality of life, mental health and life satisfaction remarkably similar to healthy controls,[111] but dismiss this as irrelevant because those locked-in patients could communicate. They take it as read that the inability to communicate makes such a dramatic difference that the condition of the patients they have in mind is likely to be much worse than death. So unless and until these patients' own views can be sought, 'terminating [their] lives might be morally *required*, not merely permissible'.[112]

Again this is too hasty. The evidence given by Laureys et al. about the quality of life of locked-in patients is consistent with well-established findings about the 'disability paradox': people with severe physical impairments report a very similar quality of life to the general population, but health professionals and others without impairments rate the quality of those people's lives much lower.[113] People who acquire impairments in adult life (for example through traumatic injury) report that before their injuries they rated the quality of such a life much lower than they do now that they experience it first-hand. In short, it is all too common for people looking in from the outside, including health care professionals, to judge some patients' lives not worth living when the patients themselves find their lives *well* worth living – and that negative judgement may translate into a failure to give treatment from which they would benefit.[114] Kahane and Savulescu's line of argument displays this disturbing tendency. Doubtless it is an awful situation to be paralysed and totally unable to communicate, as the evidence presented by Laureys et al. indicates. Kahane and Savulescu are quite right that this makes it a matter of urgency to develop communication

[110] Kahane and Savulescu, 'Brain Damage and the Moral Significance of Consciousness', p. 20.

[111] Steven Laureys et al., 'The Locked-In Syndrome: What Is It Like to be Conscious but Paralyzed and Voiceless?', *Progress in Brain Research* 150 (2005), pp. 495–511.

[112] Kahane and Savulescu, 'Brain Damage and the Moral Significance of Consciousness', p. 21, emphasis original.

[113] See, for example, Gary L. Albrecht and Patrick J. Devlieger, 'The Disability Paradox: High Quality of Life against All Odds', *Social Science and Medicine* 48.8 (1999), pp. 977–88.

[114] Laureys et al., 'The Locked-In Syndrome', pp. 506–8.

technologies to help such patients.[115] But they have failed to show convincingly that in the meantime, it may be a moral requirement to kill them.

Joseph Fins is similarly critical of paternalistic third-party judgements that the lives of patients with DoCs are not worth living.[116] To counter such paternalism, he maintains that patients' prior wishes should guide decisions about continuing or withdrawing life-sustaining treatment. Kahane and Savulescu also appeal to patients' prior wishes, as evidenced either by advance directives or surveys of the general public. They believe most people's prior wish would be to have their lives ended should they enter a vegetative or minimally conscious state. Yet the disability paradox and the evidence from Laureys et al. render such appeals to patients' prior wishes problematic. Many people who imagine that life with profound impairments would not be worth living, but subsequently find themselves in that condition, come to experience that life as worthwhile and satisfactory. If so, it cannot be assumed that an advance directive refusing life-sustaining treatment would be a reliable guide to the present wishes of a patient who is capable of having but not expressing wishes (Kahane and Savulescu's fourth scenario).[117] Nor can it be assumed that either advance directives or surveys of the healthy population are reliable guides to the present interests of DoC patients capable of having interests. Of course, decisions about treatment still have to be made in such difficult situations, so it may not seem entirely helpful to raise these awkward questions. Later I shall offer some further remarks about how such decisions might be made. Advance directives and other evidence of prior wishes may have a part to play, but cannot be assumed to settle the questions about patients' present wishes or interests.

Uncle Charlie revisited: Persons, consciousness and moral regard

In the last section, I tried to engage critically with the ethical arguments of Savulescu and others on their own terms. In the present section, I shall explore how a Christian theological ethic might approach the questions they raise. I shall

[115] Kahane and Savulescu, 'Brain Damage and the Moral Significance of Consciousness', p. 21.
[116] Fins, 'Border Zones of Consciousness', p. 23.
[117] So Fernández-Espejo and Owen, 'Detecting Awareness after Severe Brain Injury', p. 807.

claim that a theological engagement requires us to reframe the issues and address them in rather different ways. In particular, I shall argue that debates about the personhood of PVS patients, which play a dominant role in these debates, are less helpful in resolving the issues than is often supposed. I shall instead suggest an alternative, more satisfactory way to think theologically about these questions.

In ethical debates about DoCs, both sides use the concept of a person to define the borders of a moral community whose members have various moral claims on one another, including the right to life. Fins does so explicitly, using the metaphor of the border zone to argue that evidence of consciousness requires us to bring patients across the border and accord them membership of 'a human community marked by mutual interaction'.[118] This, he maintains, will mean recognizing their rights – on the one hand, the right to care (including rehabilitation and support with communication); on the other, the right to die: that is, the right to withdrawal of life-sustaining treatment if that accords with their present or prior wishes.[119] By contrast, Levy and Savulescu maintain that some forms of conscious awareness will generate certain moral obligations but will not justify including those individuals in the moral community of persons.[120] Moreover, even if some patients are brought into that community, the consequences for their right to life might be unexpected. If the conscious, sapient patients envisaged by Kahane and Savulescu qualify as persons, then ironically it is the very qualities which earn them their place in the moral community that – according to Kahane and Savulescu – may make it obligatory to end their lives.

Magisterial Catholic teaching also deploys a concept of persons as bearers of rights,[121] but uses very different criteria for identifying persons. What makes a human being a person is the presence of a human soul.[122] Scientific evidence might give indicative reasons to discern this, but a soul's presence or absence

[118] Fins, 'Border Zones of Consciousness', p. 52.

[119] Ibid.; Joseph J. Fins, 'Clinical Pragmatism and the Care of Brain Damaged Patients: Toward a Palliative Neuroethics for Disorders of Consciousness', *Progress in Brain Research* 150 (2005), pp. 565–81 (576–80).

[120] Levy and Savulescu, 'Moral Significance of Phenomenal Consciousness', pp. 367–68.

[121] A patient in PVS is still a human person with 'intrinsic value and personal dignity', and as such has rights that include 'the right to basic health care': John Paul II, 'Address to the International Congress on Life-Sustaining Treatments', paras. 3, 4.

[122] John Paul II, Encyclical Letter *Evangelium Vitae* (25 March 1995), §60. Online at http://w2.vatican.va/content/john-paul-ii/en/encyclicals/documents/hf_jp-ii_enc_25031995_evangelium-vitae.html (accessed 13 June 2016). Note that the concept of the soul in play here is much closer to a Thomist view than to the substance dualism criticized earlier in the chapter. For commentary on this aspect of John Paul II's thought and its significance for DoCs, see Pia Matthews, *Pope John Paul II and the Apparently 'Non-Acting' Person* (Leominster: Gracewing, 2013), especially chs. 2, 3, 9.

cannot be proved empirically. In particular, attributes like consciousness or the 'ability to conceive of oneself as a being persisting through time, to recall one's past, to plan, and to have preferences for how one's life goes'[123] are neither necessary conditions for an individual to count as a person, nor appropriate criteria by which to judge whether or not she is. This view will resist any separation of *human being* from *person*: 'how could a human individual not be a human person?'[124]

The fact that the concept of personhood can be used in such different ways might suggest that it is ill-suited to settling contentious issues about the moral status of human beings at the margins of life. If so, recalling its history could indicate why. From its origins in classical drama, it was initial co-opted by Christians as a load-bearing concept (so to say) in the context of Trinitarian doctrine and Christology: to speak of the three persons of the Godhead, and the two natures of Christ in one person. Indeed, the well-known definition by Boethius, 'an individual substance of a rational nature', originated in a Christological argument, though it was designed to be applicable to divine, angelic or human persons.[125] Boethius' definition remained standard in the mediaeval West,[126] but it took some significant reworkings of the concept in modernity to pave the way for its characteristic contemporary uses in bioethics. Specifically, John Locke's definition, 'a thinking intelligent Being, that has reason and reflection, and can consider itself as itself, the same thinking thing in different times and places',[127] detached the concept from Boethius' substance metaphysics and foregrounded two of the features prominent

[123] Levy and Savulescu, 'Moral Significance of Phenomenal Consciousness', p. 367.

[124] Congregation for the Doctrine of the Faith, 'Instruction on Respect for Human Life in Its Origin and on the Dignity of Procreation *Donum Vitae*' (22 February 1987), I, No. 1, quoted by John Paul II, *Evangelium Vitae*, §60.

[125] Boethius, *Theological Tractates* (E. Capps, T. E. Page and W. H. D. Rouse (ed.); H. F. Stewart and E. K. Rand (trans.); Loeb Classical Library; London: William Heinemann, n.d.), Tractate 5, ch. 3. Online at http://www.ccel.org/ccel/boethius/tracts (accessed 5 November 2016).

[126] For example, Thomas Aquinas, *The Summa Theologiae of St. Thomas Aquinas* (Kevin Knight (ed.); Fathers of the English Dominican Province (trans.); 2016) 1a, q. 29, art. 1. Online at http://newadvent.org/summa/index.html (accessed 5 November 2016).

[127] John Locke, *An Essay Concerning Human Understanding* (Peter H. Nidditch (ed.); The Clarendon Edition of the Works of John Locke; Oxford: Clarendon, 1975), 2.27.9. Ian McFarland suggests that the decisive change took place with Boethius, in whom he sees strong continuity with Locke: 'Who Is My Neighbor? The Good Samaritan as a Source for Theological Anthropology', *Modern Theology* 17.1 (2001), pp. 57–66 (58 and n. 7). But the shift I have noted, from substance to continuity of consciousness and self-consciousness, seems to indicate some fairly significant *dis*continuities. On Boethius' substantialist view, see further Peter Simpson, 'The Definition of Person: Boethius Revisited', *The New Scholasticism* 62.2 (1988), pp. 210–20.

in current debates: self-consciousness and continuity over time. (Recall the importance of these features for Levy and Savulescu's argument that signs of conscious awareness would not necessarily qualify patients as persons.) Later, the Kantian notion that the rational nature of persons is of infinite and irreplaceable value prepared the ground for the contemporary bioethical use of 'person' to designate a class of beings with distinctive value, worthy of a distinctive kind of moral regard.[128]

In short, the concept of person has been understood in such multifarious ways in Western intellectual history that it can be pressed into service to support a bewilderingly wide range of ethical positions, from Catholic Magisterial teaching to the diametrically opposite stance of Savulescu and his colleagues. Yet there may be an asymmetry about these uses, because the understandings of the concept closest to its roots in Christian thought are those that least lend themselves to the kind of practical ethical work discussed in this chapter.[129] If Christians wish to think Christianly about the moral regard due to patients with DoCs, they might find the category of personhood less well suited to the purpose than is often supposed.

This thought is reinforced by an early essay of Stanley Hauerwas with a typically arresting subtitle: 'My Uncle Charlie Is Not Much of a Person but He Is Still My Uncle Charlie.'[130] Hauerwas reflected on the divergent uses of 'person' already in use in bioethics by the mid-1970s. On one side was Paul Ramsey's deontological view of persons as inviolable moral subjects, which sets limits on the ways they may be treated – for example, in biomedical research.[131] On the other were uses of personhood to mark the boundaries of distinctive moral regard; some humans on the margins of life, such as foetuses or those nearing the end of their lives, might be placed beyond the bounds.[132] Taking up a theme familiar in his writing on medical ethics, Hauerwas argued that these

[128] See further Linda Zagzebski, 'The Uniqueness of Persons', *Journal of Religious Ethics* 29.3 (2001), pp. 401–23.

[129] So McFarland, 'Who Is My Neighbor?'

[130] Stanley Hauerwas, 'Must a Patient Be a Person to Be a Patient? Or, My Uncle Charlie Is Not Much of a Person but He Is Still My Uncle Charlie', *Connecticut Medicine* 39.12 (1975), pp. 815–17. Reprinted, for example, in *Journal of Religion, Disability and Health* 8.3–4 (2005), pp. 113–19. Subsequent citations are from the latter version.

[131] Paul Ramsey, *The Patient as Person: Explorations in Medical Ethics* (New Haven, CT: Yale University Press, 1970).

[132] For example, Fletcher, 'Four Indicators of Humanhood'; Warren, 'On the Moral and Legal Status of Abortion'.

attempts to rely on 'person' as a 'regulative notion' were a sign that medical ethics was in trouble, because this regulative concept was being expected to do the moral work that 'only a substantive community and story [could] do'.[133] In other words, as he has often argued, intractable bioethical disputes were a sign of wider moral fragmentation and incoherence in society.

A generation later, Hauerwas' diagnosis seems prescient, though his prescription – to let medical practice become as fragmented as the wider society[134] – may commend itself less, for reasons to which I shall return. But if the concept of personhood is so ill-suited to Christian reflection on the proper care of those with disorders of consciousness, what might be a better approach to that reflection?

Various authors have noted the analogy between the modern bioethicist's (or indeed neuroethicist's) question, 'Who is a person?', and the lawyer's question to Jesus, 'Who is my neighbour?' (Lk. 10.29).[135] The question follows the exchange also reported in Matthew and Mark, identifying the law (or its greatest commandments) with love of God and neighbour. But in Luke's version, '[w]anting to justify himself', the lawyer follows up this exchange with a supplementary question: 'And who is my neighbour?' Some interpreters read that question as an attempt to set limits on those whom the lawyer is obliged to love as neighbours; an understandable effort to keep the obligation within manageable bounds. But Jesus does not co-operate with this boundary-setting project. His reply, the story of the Good Samaritan, gives the lawyer no criteria for distinguishing neighbours from non-neighbours. Instead, it turns the question back onto him: 'Which of these three, do you think, was a neighbour to the man who fell into the hands of the robbers?' (v. 36). The implication is that the question, 'Who is (or is not) a neighbour whom I am obliged to love?' is the wrong one for the lawyer to ask. The question Jesus invites him to ask himself is more like, 'How am I called to *be* a neighbour to those I encounter along the way?' And the nearest thing to concrete guidance offered by the story

[133] Hauerwas, 'Must a Patient Be a Person?', p. 115.

[134] Ibid., p. 118.

[135] For example, Richard B. Hays, *The Moral Vision of the New Testament: Community, Cross and New Creation* (New York: HarperCollins, 1996), p. 451; McFarland, 'Who Is My Neighbor?'; Oliver O'Donovan, 'Again: Who Is a Person?', in M. Therese Lysaught and Joseph J. Kotva, Jr, with Stephen E. Lammers and Allen Verhey (eds.), *On Moral Medicine: Theological Perspectives in Medical Ethics* (Grand Rapids, MI: Eerdmans, 3rd edn, 2012), pp. 367–71. For a discussion of this line of reasoning in relation to a different set of ethical issues, see Neil Messer, *Respecting Life: Theology and Bioethics* (London: SCM, 2011), ch. 4.

is that being a neighbour means showing 'mercy' or compassion; 'Go and do likewise', says Jesus (v. 37).

By analogy, faced with fraught ethical choices about the care of patients with DoCs, the question 'Who is a person?' may be the wrong one – or at any rate, not the most illuminating or helpful – to ask. We might be best advised not to try and discern what is right by looking for a boundary between persons and non-persons, whether that boundary is marked out more inclusively (as by Fins) or exclusively (as by Levy and Savulescu).

If this is correct, then it turns out that for Christian ethics also, the ethical significance of consciousness is surprisingly different from what is often supposed in these debates – but what follows from this is almost opposite to the conclusions drawn by Kahane, Levy and Savulescu. Consciousness no longer has to serve as a boundary marker defining the limits of personhood and the right to life. If the right question is, instead, 'What does it mean to be a neighbour – one who acts with love and compassion – to a patient with a DoC?', then neuroimaging evidence of brain activity and consciousness may assume a more modest, but nonetheless invaluable, role in helping to inform a concrete answer to that question. What might this mean in practice?

Traumatic brain injuries on the Jericho road: Some practical conclusions

In the first century, if the robbers on the road to Jericho had inflicted a severe enough brain injury on the man in the story, he would simply have died, and all the Samaritan's oil, wine and bandages would have been to no avail. But in the twenty-first century, someone who suffers a similar assault may survive in a vegetative or minimally conscious state. So what might it mean to act as neighbours, with love and mercy, to those who suffer disorders of consciousness on our twenty-first- century equivalents of the Jericho road?

First, this ethical perspective will be strongly critical of third-party judgements that the lives of patients with DoCs are not worth living. This critical stance will be informed partly by my earlier objections to Kahane, Levy and Savulescu's arguments. Beyond those disputes, however, it is informed by the conviction that the distinctive value of any human life does not consist in self-awareness, continuity of consciousness, or the capacity to have plans, preferences, relationships or even interests. Nor, I would add, is its value best expressed by insisting that every human being is a *person* regardless of her capacities or potential. We might say that at its most basic, the value of all our lives originates in our being known and loved by God.[136] Any anxiety or frustration about the failure of some lives to fulfil human potentialities is therefore out of place: first, because this theological perspective simply denies that human worth is something we have to achieve, and second, because our hope is not 'for this life only' (1 Cor. 15.19). For any of us, whatever our abilities or impairments, the complete fulfilment of the human lives God calls us to live is an eschatological promise dependent on divine grace.[137]

We are of value because God loves us and calls us by name, a love and calling seen most fully in the life, death and resurrection of Jesus Christ. And we are called to love our neighbours in a similarly unconditional manner: to 'go and do likewise' (Lk. 10.39).[138] But what this requires in particular cases will vary, because patients with DoCs and their needs are diverse. This is one of the important lessons that Fins and others learn from neuroimaging studies: that the generalized diagnostic categories of VS and MCS, based on clinical observation alone – valuable though they have been in their way – conceal a diverse range of pathologies with varied possibilities for recovery, and an equally diverse range of needs.[139] This suggests that we should be very

[136] Warren Brown, whom I criticized in the course of my earlier discussion of the soul, affirms that 'relatedness to God would be the primary source of dignity as a value attribution, and of "soul" as a functional property of human nature': Brown, 'Neurobiological Embodiment of Spirituality and Soul', p. 71. Yet as I argued above (pp. 120–1), by identifying 'soulishness' with some level of *capacity* (however limited) for relatedness, his account leaves the door open to some conclusions about the value of human life that – I have claimed – Christians should resist.

[137] Cf., in relation to a related area of ethical concern, Reinders, *Receiving the Gift of Friendship*. More recently Reinders has expressed reservations about his earlier eschatological perspective, on the grounds that it seems to leave no space for an account of people with profound disabilities as God's *creatures*: Reinders, 'Understanding Humanity and Disability: Probing an Ecological Perspective', *Studies in Christian Ethics* 26.1 (2013), pp. 37–49. Acknowledging this concern, it nonetheless seems to me that we cannot do without *some* account of the hope of eschatological fulfilment in speaking theologically of the worth of human lives, including those with profound impairments.

[138] As McFarland notes, for much of Christian history, exegesis of the Good Samaritan story has identified the Samaritan with Jesus: 'Who Is My Neighbor?', pp. 62–63. If so, the parable's call to be neighbours by showing love and mercy is quite properly understood as a call to a kind of *imitatio Christi*.

[139] For example, Fins, 'Neuroethics and Neuroimaging', pp. 46–47.

wary of making ethical generalizations about how patients with DoCs (as a homogeneous class) should be treated – or, for that matter, dividing them for ethical purposes into two classes labelled 'VS' and 'MCS', or even a fourfold grid according to their 'consciousness' and 'sapience'.[140] There are of course some generalizations that must be made, because there are ways in which no human creature should ever be treated; but that does little more than mark the outer limits of ethical response. A key insight from Hauerwas, McFarland and others is that it is a mistake to think we can discern our essential obligations to others by getting them into the right category (such as 'person') and figuring out how members of that category in general should be treated. A patient with a DoC, like anyone else, is always a *particular* person, with a particular network and history of relationships. She or he is someone's daughter, or son, or partner – or, for that matter, Uncle Charlie.

In this connection, it is noteworthy how, in many of Joseph Fins' patient narratives,[141] it was family members who first discerned signs of patients' awareness, and advocated tenaciously for them to receive appropriate care and rehabilitation. According to Fins, such advocacy often faces severe resistance from health care professionals who see family members' claims as denial or wishful thinking. This is understandable enough, not least because denial and wishful thinking are genuine hazards of loving relationships when families are faced with the terrible reality that a loved one has a DoC. But Fins – himself a clinician with experience at the forefront of this work – argues that various features of their professional culture and context may lead clinicians to disregard, with potentially tragic consequences, *genuine* signs that family members have discerned in their loved ones.

One point that follows from this is the pressing need for a better understanding of DoCs, *inter alia* to enable more accurate and fine-grained diagnosis and identification of particular patients' needs, develop better prognosis based on a fuller understanding of the ways their conditions might evolve and identify promising approaches to rehabilitation or therapy. For this reason, Fins and others call for the 'enfranchising' of patients with DoCs in clinical research. Although such patients lack the capacity to consent to being research subjects, Fins et al. see their involvement in research as a matter of justice,

[140] *Pace* Kahane and Savulescu, 'Brain Damage and the Moral Significance of Consciousness'.
[141] Fins, *Rights Come to Mind.*

consistent with the spirit of ethical guidelines such as the US Government's *Belmont Report*.[142] Others are much warier, for two main reasons.[143] One is that it is widely considered unethical to involve human subjects incapable of informed consent in research with more than minimal risk or burden that is not intended for their direct benefit.[144] The other is the danger (acknowledged by Fins et al.) of promoting a 'therapeutic misconception' that the research will directly benefit the subject.[145] This could encourage false hope in patients' family members and induce them to give proxy consent under the mistaken impression that their loved ones would benefit.

Perhaps, though, these objections rest on an over-rigorous interpretation of benefit and therapeutic misconception. If, as Fins argues, patients with DoCs are often incompletely diagnosed and their conditions poorly investigated, then the detailed investigations involved in a research study could offer them, as well as others, benefits – particularly, as Fernández-Espejo and Owen argue, if results relevant to their clinical care are disclosed to their family members.[146] Of course it is very important not to raise false hopes or collude with family members' denial of their loved one's condition, but the answer to that concern surely lies mostly in open and careful communication.

Likewise, the informed consent requirement in research arose as a response to the notorious abuses of vulnerable populations as research subjects during and after the Second World War.[147] It remains a crucial safeguard, all too easily eroded by powerful commercial or other interests. Yet it would seem perverse if rigid adherence to it, in isolation from the context of DoC patients' lives and needs, were to prevent the acquisition of knowledge necessary to benefit members of this patient population – maybe including the research subjects themselves, as Fernández-Espejo and Owen argue. It goes without saying that

[142] For example, Fins et al., 'Neuroimaging and Disorders of Consciousness', pp. 8–9, citing National Commission for the Protection of Human Subjects, *The Belmont Report: Ethical Principles and Guidelines for the Protection of Human Subjects of Research* (DHEW Publication No. (OS) 78-0012; Washington, DC: U.S. Government Printing Office, 1978).

[143] For example, Schwartz and Schwartz, 'The Risks of Reducing Consciousness to Neuroimaging'.

[144] Indeed, this was a major preoccupation of Ramsey, *The Patient as Person*. An often-cited example of research with greater than minimal risk or burden would be to investigate DoC patients' neural responses to unpleasant or mildly painful stimuli.

[145] See Paul S. Appelbaum, Loren H. Roth and Charles Lidz, 'The Therapeutic Misconception: Informed Consent in Psychiatric Research', *International Journal of Law and Psychiatry* 5.3–4 (1982), pp. 319–29.

[146] Fernández-Espejo and Owen, 'Detecting Awareness after Severe Brain Injury', pp. 807–8.

[147] See Albert R. Jonsen, *The Birth of Bioethics* (Oxford: Oxford University Press, 1998), ch. 5.

careful proxy consent procedures and control of risks and burdens, as outlined by Fins et al., would be *sine qua non*.[148]

A related point emphasized by Fins is the need for advocacy on behalf of this patient group to challenge 'nihilistic' professional attitudes that dismiss the possibility of improvement in patients' conditions, and the contextual and systemic barriers hindering access to care and rehabilitation from which they could benefit.[149] The theological perspective I have outlined will certainly support this call for advocacy, though with two reservations. One I have already stated: I am not convinced that asserting the *personhood* of (some of) these patients will offer as secure a basis for advocacy as Fins thinks. The other is that, understandably in his context, he adopts a model of advocacy based on the disability rights movement.[150] As such, from a theological perspective his approach will have both the strengths and the limitations of the civil rights model in disability advocacy. The strengths are too evident to need spelling out. But as various theological critics have argued, the civil rights paradigm assumes that the goal of advocacy is participation in society as active self-determining individuals, and this is a problematic aim for some – including those with profound cognitive impairments.[151] From a theological perspective, advocacy is needed not only for those patients who may be able to regain some measure of active participation in society, but also for those who realistically have no such prospect; and they might not be so well served by a civil rights paradigm.

What of these latter patients, including those who really are in a permanent vegetative state, with no conscious awareness nor any realistic prospect of regaining it? As I have repeatedly emphasized, the value of any human creature does not depend on his or her capacities or potentialities. We are called to be neighbours to truly PVS patients, as to any human creature, by showing love and mercy. Much of what this requires, I have also suggested, must be discerned in each individual relationship. In general terms, though, it will involve caring for their bodily needs, including protecting them as far as

[148] Fins et al., 'Neuroimaging and Disorders of Consciousness', p. 8.

[149] Fins, *Rights Come to Mind*. Some of the contextual factors he addresses are specific to the United States and its insurance-based system of health care provision. But while the contextual factors may differ in other places, it is not only in the USA that patients with DoCs need advocacy.

[150] Ibid., ch. 22.

[151] So, for example, Reinders, 'Understanding Humanity and Disability', pp. 41–42.

possible from any residual pain or discomfort they might be capable of feeling. It will not necessarily involve prolonging their lives by all possible means, because that would be to idolize this mortal life and neglect the hope of eternal life.[152] To repeat, what this means in practice will be a matter for discernment in particular situations, which will require good and open communication between family members and health care professionals.[153] In general terms, it may mean that aggressive resuscitation or treatment of life-threatening infections might not be appropriate.

The vexed question that arises at this point, of course, is whether ANH should be considered as life-sustaining treatment that may legitimately be withdrawn. Nutrition and hydration are clearly among the most basic necessities required to sustain an embodied human life. If one is responsible for the care of someone who cannot care for themselves, then providing food and water must surely be part of that responsibility; and it is implicit in what I have said hitherto that a patient's claim on the care necessary to sustain life is not dependant on his or her capacities (or lack of them). Moreover, it is clear enough that often, when the withdrawal of ANH from PVS patients is advocated, those who advocate it have the aim of the patients' (bodily) death in view. In the theological perspective developed in this book, it is difficult to see how this could be morally justified: I am yet to be persuaded that acting as a neighbour by showing love and mercy to someone ever entails killing her.[154]

However, there might be situations in which the artificial administration of food and water is disproportionately burdensome to a PVS patient, and in those situations the withdrawal of ANH could be a morally appropriate expression of care for that patient. If that is so, it may be right for Christians to support legislation that permits it,[155] even if such legislation also makes possible more morally dubious instances of withdrawing ANH. The law is a blunt instrument for enforcing nuanced moral judgements, and in this case perhaps, the best it can do is to keep open a space in which family members

[152] See further Neil Messer, *Flourishing: Health, Disease and Bioethics in Theological Perspective* (Grand Rapids, MI: Eerdmans, 2013), pp. 180–82, 194–200.

[153] The narratives recorded by Fins include numerous stories of communication and decision-making that, to put it mildly, did not live up to this standard: for example, *Rights Come to Mind*, chs. 3, 6, 9.

[154] Some of my reasons for being unpersuaded are outlined in Messer, *Respecting Life*, ch. 8.

[155] As is the case, for example, in the UK following the Law Lords' judgement in *Bland*.

and health care professionals are able to exercise their moral responsibility before God for patients in their care.[156]

At the other end of the scale are those patients who may meet behavioural criteria for VS or MCS, but who seem able to use mental imagery detectable by neuroimaging to communicate and answer questions. It is widely agreed that there is a pressing need to develop technologies, such as neuroimaging or brain–computer interfaces, that will help these patients communicate as effectively as possible. The good of such technologies is obvious: they would restore patients to fuller and more active relationships with families and loved ones, would enable them to communicate their needs to carers, and may enable them to express their own wishes about their treatment (and for that matter, participation in research).[157] If such patients were enabled to communicate, one possible result would be that they expressed a wish to die. If surveys of locked-in patients and the lessons of the disability paradox are anything to go by, many would not; but some might.[158] In which case, how should family members and health care teams respond?

Here distinctions often elided in such debates become important. Anyone with the necessary mental capacity may refuse life-sustaining treatment. It might be quite appropriate for carers and family members to enter into dialogue and interrogate that refusal, so far as circumstances allowed, but if it were a firmly held decision it should be respected. It would then be the business of the health care team to make the final stages of the patient's life as comfortable and free of pain and distress as possible. A request for active euthanasia would be a different matter, and there are theological reasons to think that such a request need not and should not be agreed to.[159] A request for the withdrawal of ANH occupies a difficult middle ground. It follows from my earlier discussion that there might be circumstances in which patients could legitimately judge ANH to be disproportionately burdensome to them, and therefore request its withdrawal. In any event, it would be a denial of their

[156] Cf. Karl Barth, *Church Dogmatics*, vol. 3.4, p. 422.

[157] For significant decisions concerning their health care or participation in research, it would also be necessary to find ways to assess their mental capacity for decision-making, which is not necessarily coterminous with the ability to communicate and answer questions.

[158] It would therefore be quite unjustified to assume, in the absence of any clear indication either way, that this *is* their wish, and act accordingly; *contra* Kahane and Savulescu, 'Brain Damage and the Moral Significance of Consciousness', pp. 19–21.

[159] See further Messer, *Respecting Life*, ch. 8.

freedom and responsibility as creatures before God to force them to receive ANH if they were determined to refuse it. There are reasons to think, therefore, that patients' competent *present* refusal of ANH should be respected.

I emphasize '*present* refusal', because the question of advance directives arises once again. As I argued earlier, they should be treated with great caution as guides to decision-making where there are indications that patients might have present wishes or interests, because we do not have good grounds to think that advance directives made when their authors *imagined* having a DoC are reliable indicators of their experience or wishes once they are in such a condition. However, when it is clear that a patient is incapable of having present wishes or interests – when she is truly in a PVS – then an advance directive could serve as one aid among others, guiding those who have to make a responsible decision about what is entailed by showing her love and mercy. It should be treated in that way: as an aid to decision-making by those with that responsibility, not as something akin to an autonomous expression of the patient's will that should override other considerations. For that reason, there will be limits to what an advance directive can legitimately direct. My earlier argument suggests that a good deal of caution would be in order about implementing an advance directive requesting withdrawal of ANH, though there could be circumstances and reasons that would render such a decision morally appropriate.

I noted earlier that resource allocation arguments are sometimes given as reasons to end the lives of patients with DoCs. Dominic Wilkinson et al. put it like this: 'considerations of distributive justice may warrant withdrawal of treatment when such treatment (for example intensive care) has finite availability and other individuals have greater prospects of recovery'.[160] It is true that specialist care and rehabilitation of DoC patients are costly and wide use of new technologies like neuroimaging or BCIs would add to the costs. Some of Fins' patient narratives show how, in an American context of insurance funding, decisions about patient care, discharge and access to rehabilitation can be quite brutally driven by resource considerations.[161] But resource pressures are hardly unique to that context or system, even if the ways they play out are highly context-specific. And even if money is not scarce,

[160] Wilkinson et al., 'Functional Neuroimaging and Withdrawal of Life-Sustaining Treatment', p. 509.
[161] Fins, *Rights Come to Mind*, ch. 9.

some other resource – such as specialist facilities or professional expertise – may be, as Fins also makes clear.[162]

Resource allocation questions are real and difficult; but the remark of Wilkinson et al. suggests that other patients with 'greater prospects of recovery' have a stronger claim on the resources of the health care system than patients in VS or MCS. And this appears to presuppose either a ranking of the value of patients' lives according to their capacities and potential, or a utilitarian approach along the lines of the quality-adjusted life year (QALY) system, or both. I have already explained why the theological ethic articulated in this chapter should resist any ranking of the value of patients' lives, and such an ethic should also make us deeply suspicious of QALY-based approaches.[163] Difficult resource allocation questions will have to be faced in any system, and may sometimes necessitate tragic choices, but we must find ways to make those choices without resorting to questionable judgements about the worth or quality of different patients' lives.[164]

I believe all of the practical conclusions outlined in this section are defensible answers to the question, 'What might it mean to be neighbours, showing love and mercy, to patients with disorders of consciousness?' Some of my conclusions might not look distinctively theological, and might be shared by many people who do not approach these questions theologically. That in itself is not a problem: the calling of Christian ethics is not to be distinctive for distinctiveness' sake, but to be faithful, and to be distinctive when faithfulness requires it. It might even be an advantage, in that it will offer opportunities to make common cause with others in pursuit of agreed goods and goals. Others of my conclusions, however, are more peculiarly theological. The refusal to ground the value of human lives in their capacities or potentialities, for example, will seem odd to many.

Four decades ago Hauerwas, having argued that appeals to personhood in bioethics were a symptom of deep moral fragmentation, suggested that 'we will be much better off to simply admit that morally there are many different ways to practice medicine. We should, in other words, be willing to have our

[162] Ibid., for example, p. 105.

[163] Reasons for such suspicion are given in Messer, *Respecting Life*, ch. 7, and idem, *Flourishing*, pp. 208–9.

[164] For a sketch of how such questions might be approached theologically, see Messer, *Respecting Life* ch. 7.

medicine as fragmented as our moral lives'.[165] If this should be taken to mean not attempting to influence the health care system as a whole in the directions I have outlined, then I disagree. What such a conclusion would fail to recognize is that Christians might have theological reasons to take responsibility for how things go in the world outside the Church. In Dietrich Bonhoeffer's language, the world is, by God's grace, the 'penultimate' realm in which people may encounter the 'ultimate' – God's word of judgement and salvation in Christ. There is therefore a theological mandate for Christians to be concerned for all areas of life in the world, and to do what we can to make them fit to be penultimate arenas. One can agree entirely with Hauerwas about the moral fragmentation of contemporary societies and their health care systems, but still consider that a state of affairs in which patients' capacities do not determine their worth or their care is more truly penultimate than other scenarios. If so, Christians have good theological reasons to try and make such a state of affairs a reality using any legitimate means available – not only the lived witness of the Church, but also (for example) argument, advocacy and alliance-building with others. All these and more could be proper responses to the command to 'go and do likewise'.

[165] Hauerwas, 'Must a Patient Be a Person?', p. 118.

6

Messing with Our Minds: The Ethics of Technological Interventions in the Brain

Introduction

Neuroscience, like other biomedical sciences, is a practical as well as a theoretical discipline. This was clear back in 1990 when then-US President George Bush, Sr., proclaimed the 'Decade of the Brain': most of the emphasis in his presidential proclamation was on understanding 'what goes wrong when [the brain] is injured or diseased' and how new treatments for brain damage and disorders might be developed.[1] Researchers who study the human brain are not simply seeking knowledge of how the brain works for knowledge's sake; often they also hope to find new ways to *modify* its working.[2] Indeed, the decades since the Decade of the Brain have not only brought great advances in understanding how the brain works, but also new and growing possibilities for intervening technologically in its functioning. Many of these neurotechnologies have been developed initially to treat psychiatric or neurological disorders. But some of the same technologies are also used in an attempt to enhance the capacities and functions of healthy brains, and some authors anticipate much more powerful and far-reaching neural enhancements in the future.

This chapter is concerned with the ethics of such technological interventions in the brain. First, I shall survey some of the neurotechnologies that are either

[1] George Bush, Presidential Proclamation 6158 (17 July 1990), online at http://www.loc.gov/loc/brain/proclaim.html (accessed 13 March 2017).

[2] As has long been recognized, this drive to understand the world in order to control it for the benefit of humanity has been at the heart of modern science and technology since their beginnings; this has also given rise to some of the moral ambiguities and tensions associated with the practice of science and technology, including those which will be explored in this chapter. For a well-known and influential analysis, see Gerald P. McKenny, *To Relieve the Human Condition: Bioethics, Technology, and the Body* (Albany: State University of New York Press, 1997), pp. 17–24.

currently available or anticipated for the future. Next, I shall offer a critical survey of philosophical debates about the ethics of cognitive enhancement, focusing on Michael Sandel's critique of enhancement and responses from some of his critics, such as John Harris. Finally, I shall develop a theological perspective on these questions. I shall argue (against philosophers such as Harris) that a distinction between therapy and enhancement is both coherent and ethically important. I shall propose a way to locate the ethical evaluation of these practices in the context of the core themes of the Christian faith, and argue that an ethical evaluation of cognitive enhancement depends on a theological understanding of the practices (such as education) that it is supposed to enhance.

Examples of interventions: Current practice and future speculation

Pharmacological cognitive enhancements

Humans have used chemical substances to alter mental functioning and mood for millennia: the chewing of coca leaves began in South America at least 8,000 years ago,[3] while the authors of Hebrew biblical texts were well aware that alcohol could 'gladden the human heart' (Ps. 104:15), but might also have less welcome side effects (Prov. 23:29-35). In modern times, psychiatry has been transformed since the 1950s by successive generations of antipsychotics, antidepressants, stimulants and other groups of psychotropic drugs. The rate of new psychiatric drug development is however said to be slowing for various reasons;[4] this has prompted renewed interest in non-pharmaceutical approaches to treating mental and neurological disorders, including some of the novel neurotechnologies outlined below.

In addition to their use to treat recognized mental disorders or clinical symptoms, some of these drugs are used off-label to enhance mental functions. For example, among the stimulants, modafinil is used clinically to treat

[3] Tom D. Dillehay et al., 'Early Holocene Coca Chewing in Northern Peru', *Antiquity* 84.326 (2010), pp. 939–53.

[4] Nuffield Council on Bioethics, *Novel Neurotechnologies: Intervening in the Brain* (London: Nuffield Council on Bioethics, 2013), p. 4.

narcolepsy, and methylphenidate (better known as Ritalin) is widely used for attention deficit hyperactivity disorder (ADHD), particularly in children and adolescents. Both are also used by people without these diagnoses to enhance learning, memory, attentiveness and other cognitive functions. Their use by students aiming to improve their grades has attracted particular academic and media attention.[5] Estimates of the prevalence of such use vary widely. Most data are from the United States, where in different studies anything from 5 to 35 per cent of college students report having used stimulants for non-medical reasons. Prevalence has been found to vary between institutions and student populations: in particular, male students, those at competitive institutions, and members of fraternity and sorority societies are more likely to report such stimulant use.[6]

Less evidence is available outside the United States, but this kind of stimulant use appears to be less prevalent. In one recent study of students in UK and Irish universities, under 10 per cent reported having used stimulants for cognitive enhancement, and 3.4 per cent were currently using modafinil, the most commonly used of the drugs surveyed. Users were most likely to be white, British and male, and either postgraduates or undergraduates near the end of their courses. Peer pressure seemed to be an important factor in encouraging use.[7] Apart from students, reports suggest that other groups, including academics, health care professionals and military personnel, also use these stimulants.[8]

There is mixed evidence for the efficacy of these drugs as cognitive enhancers,[9] and it has been claimed that non-pharmacological methods – both time-honoured ones like sleep and exercise and new neurotechnologies of

5 For reviews of the academic literature, see, for example, M. Elizabeth Smith and Martha J. Farah, 'Are Prescription Stimulants "Smart Pills"? The Epidemiology and Cognitive Neuroscience of Prescription Stimulant Use by Normal Healthy Individuals', *Psychological Bulletin* 137.5 (2011), pp. 717–41, and C. Ian Ragan, Imre Bard and Ilina Singh, 'What Should We Do about Student Use of Cognitive Enhancers? An Analysis of Current Evidence', *Neuropharmacology* 64 (2013), pp. 588–95. For an example of British media coverage, see Carole Cadwalladr, 'Students Used to Take Drugs to Get High. Now They Take Them to Get Higher Grades', *The Observer* (15 February 2015), online at http://www.theguardian.com/society/2015/feb/15/students-smart-drugs-higher-grades-aderall-modafinil (accessed 7 April 2016).

6 Thomas Metzinger and Elisabeth Hildt, 'Cognitive Enhancement', in Judy Illes and Barbara J. Sahakian (eds.), *The Oxford Handbook of Neuroethics* (Oxford: Oxford University Press, 2011), pp. 245–64 (249); Smith and Farah, 'Are Prescription Stimulants "Smart Pills"?'.

7 Ilina Singh, Imre Bard and Jonathan Jackson, 'Robust Resilience and Substantial Interest: A Survey of Pharmacological Cognitive Enhancement among University Students in the UK and Ireland', *PLoS ONE* 9.10 (2014), doi: 10.1371/journal.pone.0105969.

8 Barbara Sahakian and Sharon Morein-Zamir, 'Professor's Little Helper', *Nature* 450 (2007), pp. 1157–59; Brendan Maher, 'Poll Results: Look Who's Doping', *Nature* 452.7188 (2008), pp. 674–75; Metzinger and Hildt, 'Cognitive Enhancement', pp. 249, 258–59.

9 For a review, see Smith and Farah, 'Are Prescription Stimulants "Smart Pills"?'.

the sort outlined below – may be more effective than drugs.[10] Moreover, it is not clear that everyone would benefit, or benefit equally, from stimulants like modafinil and methylphenidate. Some cognitive enhancing effects seem to be baseline-dependent: pronounced in people with lower baseline levels of ability, but less marked or absent in those with higher baseline abilities.[11] Nonetheless, most authors assume that these drugs have some efficacy in cognitive enhancement, and that there will be more effective ones in the future.

Novel neurotechnologies

Like psychopharmacology, other forms of intervention such as surgery and electrical stimulation have a history going back to the early or mid-twentieth century.[12] That history has some dark episodes – particularly the use of lobotomy, which fell into disrepute as the evidence accumulated that it left many patients with debilitating brain damage. However, more careful and less damaging techniques were also developed during the twentieth century, including precursors of technologies in use or under development today.

These new approaches are diverse, but a common feature is that they all involve direct interventions in the brain. *Transcranial brain stimulation* (TBS) is a group of non-invasive approaches that use either a magnetic field or an electric current (delivered via scalp electrodes) to interfere temporarily with the electrical activity in the targeted part of the brain. It is widely used in neuroscientific research, and is beginning to be applied therapeutically. Transcranial magnetic stimulation (TMS) in particular has shown some success in treating drug-resistant depression. Applications for many other psychiatric and neurological disorders are being researched, but are less well developed than its use in depression.[13]

Unlike TBS, *deep brain stimulation* (DBS) is an invasive technique, in which fine electrodes are inserted through the skull into a subcortical nucleus of the brain (the exact location depends on the condition being treated).

[10] Martin Dresler et al., 'Non-Pharmacological Cognitive Enhancement', *Neuropharmacology* 64 (2013), pp. 529–43.

[11] Reinoud de Jongh et al., 'Botox for the Brain: Enhancement of Cognition, Mood and Pro-Social Behavior and Blunting of Unwanted Memories', *Neuroscience and Biobehavioral Reviews* 32 (2008), pp. 760–76.

[12] Nuffield Council, *Novel Neurotechnologies*, pp. 3–4.

[13] Ibid., pp. 18–19.

Pulses of electric current are applied to the target area via these electrodes. One of its most successful applications is in treating some of the symptoms of Parkinson's disease, particularly in cases resistant to drug therapy. This success has prompted other applications for conditions such as chronic pain, epilepsy, migraine and Alzheimer's disease (though with varying degrees of success) and there is interest in using it for treatment-resistant cases of psychiatric disorders like depression and obsessive-compulsive disorder.[14] The mechanisms of action of TBS and DBS are not well understood at present.[15]

Brain–computer interfaces (BCIs) connect individuals' brains to external devices.[16] Signals from the brain are picked up by electrodes attached to the scalp, implanted inside the skull or inserted into the brain. A computer program converts the signals into information that is used to control a device. With training, the user can learn to direct the device by means of her thoughts. At the time of writing, BCI applications are still at the research stage, with interest focused on three areas. The most advanced is the development of assistive technologies (such as robotic arms and wheelchairs) for people paralysed or 'locked-in' as a result of brain injury or disease. A second is neurorehabilitation, where BCIs may help regain motor function that has been impaired by disease or injury. A third, as we saw in Chapter 5, is the use of BCIs to test for awareness in patients unable to communicate, to determine whether they are in locked-in, minimally conscious or vegetative states.

A fourth group of novel neurotechnologies is *neural stem cell therapies*. Stem cells are relatively undifferentiated cells with the potential to differentiate into more specialized cell types. Depending on their source, they vary in potency. Embryonic stem cells have the useful property of being pluripotent (able to generate all the cell types found in the adult body), but can only be obtained by destroying the embryos they come from. No naturally occurring adult stem cells are pluripotent. However, in recent years, it has become possible to 'reprogram' differentiated adult cells. The resulting *induced pluripotent stem cells* have properties very similar to embryonic stem cells, but can be obtained without destroying human embryos. Neural stem cell therapies are directed at conditions where brain tissue is lost suddenly (as in a stroke) or gradually (as

14 Ibid., p. 24.

15 Ibid., pp. 19–21, 24–25.

16 For the following description, see ibid., pp. 28–35.

in degenerative diseases).[17] An eventual aim is to develop neuron replacement therapies (NRTs), in which the stem cells generate new neurons to replace those that have died. However, the first therapies to have reached clinical trials do not seem to work this way, but by means of a 'bystander' effect, enhancing the function of surviving neurons rather than generating new ones.[18]

As well as therapeutic applications, some of these novel technologies have possible enhancement or other non-therapeutic applications. For example, there is some evidence that under laboratory conditions TBS can enhance various cognitive functions (such as memory, language skills and reasoning), as well as emotional processing and mood, though it is not clear whether these effects would be significant in the real world.[19] Other areas of non-therapeutic use are gaming and military applications,[20] both of which are (for different reasons) beyond the scope of this chapter.

Other aspects of enhancement

Much of the interest in non-therapeutic uses of drugs and neurotechnologies focuses on *cognitive* enhancement: increasing capacities such as attention, learning and memory in individuals for whom they are not impaired by diagnosed disorders. However, cognitive capacities are not the only neural targets for enhancement. For example, antidepressant or anxiolytic drugs may be used as *mood* enhancers, and there is some discussion of using neurotechnologies such as TBS for the same purpose.[21]

Another aspect that attracts attention is *moral* enhancement. When this is mooted, it is not always clear exactly what would be enhanced and how, but Ingmar Persson and Julian Savulescu propose that the targets of moral enhancement would be moral emotions (such as sympathy, gratitude, anger, guilt, shame and forgiveness) that influence people's dispositions to altruism and justice.[22] Various drugs already used clinically (e.g., some beta-blockers,

[17] Ibid., pp. 36–40.
[18] Ibid., p. 38.
[19] Ibid., pp. 165–68.
[20] Ibid., pp. 168–69, 182–90.
[21] Ibid., pp. 164–68.
[22] Ingmar Persson and Julian Savulescu, 'The Perils of Cognitive Enhancement and the Urgent Imperative to Enhance the Moral Character of Humanity', *Journal of Applied Philosophy* 25.3 (2008), pp. 162–77 (168–69).

oral contraceptives, glucocorticoids and antidepressants) are said to have measurable effects on moral emotions and behaviour. Neil Levy et al. argue that since these drugs are widely prescribed, they might already be having unintended effects on moral behaviour at population level in ways that are little understood and require investigation.[23] In addition to these possible unintended effects, some drugs are already used with the express purpose of changing behaviour – for example, the anti-craving drugs used for alcohol dependence, discussed in Chapter 4. Beyond specific applications like the treatment of addictions, some authors are highly sceptical about the possibility of moral modification. This may be because of doubts about its technical feasibility, but others find the project of moral enhancement itself conceptually confused.[24]

More ambitiously, cognitive and other neural enhancements are one aspect of the transhumanist goal of enhancing ourselves or our descendants so radically that we (or they) become a different 'posthuman' kind of being, with capacities and ways of being that are unimaginable to present-day humans. Along the way to posthumanity would be 'transhumans', transitional beings with moderately enhanced capacities, but not the full range of posthuman possibilities.[25] Transhumanists speculate (*inter alia*) about radically enhancing intelligence, mood and self-control, developing entirely new senses, radically extending the lifespan, and uploading human minds to computers.[26] Critics however discern questionable assumptions and conceptual confusions in these speculations, especially mind uploading.[27]

23 Neil Levy et al., 'Are You Morally Modified? The Moral Effects of Widely Used Pharmaceuticals', *Philosophy, Psychiatry, and Psychology* 21.2 (2014), pp. 111–25.

24 For example, Fabrice Jotterand, 'Moral Enhancement, Neuroessentialism, and Moral Content', in Fabrice Jotterand and Veljko Dubljevic (eds.), *Cognitive Enhancement: Ethical and Policy Implications in International Perspectives* (Oxford: Oxford University Press, 2016), pp. 42–56. See also Harris Wiseman, *The Myth of the Moral Brain: The Limits of Moral Enhancement* (Cambridge, MA: MIT Press, 2016).

25 Nick Bostrom, 'Transhumanist Values', *Journal of Philosophical Research* 30, suppl. (2005), pp. 3–14 (5, fig. 1).

26 See ibid., pp. 5–7, also on life extension, Nick Bostrom, 'The Fable of the Dragon-Tyrant', *Journal of Medical Ethics* 31.5 (2005), pp. 273–77; on mind uploading, Ben Goertzel and Matthew Iklé (ed.), Special Issue on Mind Uploading, *Journal of Machine Consciousness* 4.1 (2012).

27 See Michael Hauskeller, 'My Brain, My Mind, and I: Some Philosophical Assumptions of Mind-Uploading', *Journal of Machine Consciousness* 4.1 (2012), pp. 187–200; Patrick D. Hopkins, 'Why Uploading Will Not Work, or, the Ghosts Haunting Transhumanism', *Journal of Machine Consciousness* 4.1 (2012), pp. 229–44.

Ethical debates

Therapy, including experimental therapies and new technologies

When therapeutic uses of drugs and novel neurotechnologies are considered, the ethical issues raised are mostly familiar ones such as safety, efficacy, the balance of benefits, burdens and risks, distributive justice and informed consent. While these issues are not specific to neurotechnologies, they arise in distinctive ways when interventions in the brain are being considered.[28] First, the unique properties of the human brain and the crucial part it plays in human distinctiveness, identity and personality mean that conditions affecting the brain can have a particularly devastating impact on those who suffer from them. Therefore, the need for safe and effective therapies for these conditions can seem particularly pressing. However, these same features of the brain and the limits to our present understanding of it mean that interventions to modify its function can be a hazardous undertaking, and ill-judged interventions have the potential to do devastating damage – as the history of lobotomy shows all too clearly. Moreover, some of the very conditions requiring treatment may make informed consent difficult or impossible, so questions about how decisions are made, and by whom, arise in particularly acute ways.

In response to these distinctive concerns, the UK's Nuffield Council on Bioethics has proposed a three-level ethical framework for assessing novel neurotechnologies.[29] First are two foundational principles that must be held in balance, beneficence and caution. Next, to specify the implications of these principles, the report identifies five 'interests' requiring particular attention: safety, privacy, patients' autonomy, equity of access to treatments, and trust in research and new therapies. Finally – interestingly for a report of this kind – the authors argue for a virtue-based approach to practical reasoning about particular cases. Three virtues identified as particularly relevant are inventiveness, humility and responsibility.[30]

[28] Nuffield Council, *Novel Neurotechnologies*, pp. 73–78.

[29] Ibid., ch. 4.

[30] The authors emphasize, though, that '[t]hese are intended to complement, rather than replace, the all-purpose virtues that are characteristic of almost any decent human life and society' (ibid., p. 85).

Enhancement

Many of the issues raised by therapies – such as safety, distributive justice, freedom and consent – also arise in relation to enhancement.[31] As the Nuffield Council report observes, these might play out differently for non-therapeutic than for therapeutic applications: for example, the balance of risks, harms and benefits might well be different, and it may be more difficult to monitor the safety and efficacy of these technologies when they are used outside medical contexts.[32] However, there is also a vigorous ethical debate about human enhancement that extends beyond these questions of safety, distributive justice and freedom to consider what the US President's Council on Bioethics called 'the essence of the activity itself'.[33] Much of this debate is cast in quite general terms and refers to a range of different enhancement technologies, having historically been focused more on genetic enhancements than other kinds. But many of these general arguments apply as much to neural as to other enhancements – though they may take particular forms in relation to neural enhancements, and there may also be issues and arguments more specific to interventions in the brain.[34]

One of the best-known ethical critiques of human enhancement has been developed by Michael Sandel in various publications, notably his book *The Case against Perfection*.[35] Although his principal target is genetic enhancement, Sandel emphasizes that his concerns apply equally to other forms, including the use of performance-enhancing drugs in sport and what he calls 'hyperparenting'.[36] He draws a clear – though, as he acknowledges, not entirely sharp – distinction between therapy and enhancement: the practice of medicine 'is governed, or at least guided, by the norm of restoring

31 See President's Council on Bioethics, *Beyond Therapy: Biotechnology and the Pursuit of Happiness* (Washington, DC: President's Council on Bioethics, 2003), pp. 279–85, online at https://repository.library.georgetown.edu/handle/10822/559341 (accessed 12 April 2016).

32 Nuffield Council, *Novel Neurotechnologies*, pp. 172–82.

33 President's Council, *Beyond Therapy*, p. 286. Recent collections reflecting this debate include Julian Savulescu and Nick Bostrom (eds.), *Human Enhancement* (Oxford: Oxford University Press, 2009) and Julian Savulescu, Ruud ter Meulen and Guy Kahane (eds.), *Enhancing Human Capacities* (Chichester: Wiley-Blackwell, 2011).

34 See, for example, Jotterand and Dubljevic, *Cognitive Enhancement*.

35 Michael J. Sandel, *The Case against Perfection: Ethics in the Age of Genetic Engineering* (Cambridge, MA: Belknap Press, 2007). He also made a major contribution to the somewhat similar critique developed in President's Council, *Beyond Therapy*, for example, pp. 286–95.

36 Sandel, *The Case against Perfection*, p. 52.

and preserving the natural human functions that constitute health',[37] and this means that therapy has inherently limited aims in a way that enhancement does not.

At the heart of Sandel's ethical concern is that enhancement represents a Promethean kind of 'hyperagency' or 'drive to mastery' over our own bodies and lives and those of our children.[38] The problem with this, he maintains, is that it 'represent[s] the one-sided triumph of wilfulness over giftedness, of dominion over reverence, of molding over beholding'.[39] The phrase 'one-sided' should be emphasized: Sandel is not condemning all attempts to improve nature or human life. He endorses the argument of the theologian William F. May, that in parenting a balance is needed between 'accepting' and 'transforming' love, and likewise in scientific engagement with nature we need a balance between 'beholding' the world and 'molding' it.[40] Sandel's objection to human enhancement is that it loses that balance, overemphasizing transformation at the expense of acceptance and 'molding' to the exclusion of 'beholding'. In so doing, it undermines what (in another phrase borrowed from May) he calls our 'openness to the unbidden' and our appreciation for the giftedness of human lives and abilities.[41] This, he believes, 'will transform three key features of our moral landscape – humility, responsibility, and solidarity'.[42] Both our inability to control how our children turn out, and the limits to our capacity for self-improvement, can teach us a kind of *humility* that makes us receptive to the unexpected and unbidden. The fact that human lives and abilities have this 'unbidden' element also sets limits on our *responsibility* for the way we and our children turn out, limits which may be eroded by the drive to mastery inherent in the practice of enhancement. 'The more we become masters of our

37 Ibid., p. 47.

38 Ibid., pp. 26, 27.

39 Ibid., p. 85.

40 Ibid., pp. 49–50, quoting comments by William F. May in Transcript of President's Council on Bioethics, Session 2, 'Science and the Pursuit of Perfection', 17 January 2002, online at https://bioethicsarchive.georgetown.edu/pcbe/transcripts/jan02/jan17session2.html (accessed 12 April 2016).

41 Sandel, *The Case against Perfection*, p. 45, citing comments by William F. May in Transcript of President's Council on Bioethics, Session 2, 'Choosing the Sex of Children: Demographics', 17 October 2002, online at https://bioethicsarchive.georgetown.edu/pcbe/transcripts/oct02/session2.html (accessed 12 April 2016).

42 Sandel, *The Case against Perfection*, p. 86. For the argument summarized in the remainder of this paragraph see ibid., pp. 85–100. The summary is modified from Neil Messer, *Respecting Life: Theology and Bioethics* (London: SCM Press, 2011), p. 92.

genetic endowments, the greater the burden we bear for the talents we have and the way we perform.'[43] Furthermore, Sandel argues, if it becomes routine to take control over our genetic endowments, practices of social *solidarity* by which resources are shared with those who suffer misfortune will be eroded. If, for example, children with disabilities are routinely screened out, it becomes easier to argue that those who have failed to do so should not be a drain on the resources of those who have acted more prudently.

Sandel's position has attracted no shortage of critics. Many have pursued one of two lines of criticism he anticipates: consequentialist critiques, and the objection that his argument depends on religious presuppositions and is unpersuasive to non-religious people.[44]

A somewhat knockabout example of the first can be found in John Harris' book *Enhancing Evolution*.[45] Remarking repeatedly on Sandel's rhetorical eloquence, Harris himself deploys some rather overblown rhetoric. Between the rhetorical flourishes, however, various substantive criticisms emerge. One is that the distinction between legitimate and illegitimate forms of intervention is question-begging.[46] Sandel, as we have seen, bases this distinction on the difference between therapy and enhancement, arguing that therapeutic purposes set limits on the goals pursued, thereby restraining the drive to mastery. By contrast, Harris gives the therapy-enhancement distinction short shrift earlier in his book, concluding that '[i]t does not draw either a morally significant or an explanatorily significant distinction and so fails utterly to be useful'.[47] The claim of *explanatory* insignificance rests on a series of examples which, he thinks, undermine the distinction.

[43] Sandel, *The Case against Perfection*, p. 87.

[44] Ibid., pp. 92–97. Others train their critical fire on the details of Sandel's normative argument, but in so doing perhaps miss his main point. For example, Frances Kamm treats his critique of the drive for mastery as an argument about the motives of individuals who choose enhancements, and finds it unconvincing: 'What Is and Is Not Wrong with Enhancement?', in Savulescu and Bostrom (eds.), *Human Enhancement*, pp. 91–130. But as Michael Hauskeller argues, it is better understood as a claim about the impoverishing effect of the drive for mastery on the moral character of a *community or society* in which enhancements are accepted and practiced: 'Human Enhancement and the Giftedness of Life', *Philosophical Papers* 40.1 (2011), pp. 55–79 (76–78).

[45] John Harris, *Enhancing Evolution: The Ethical Case for Making Better People* (Princeton, NJ: Princeton University Press, 2007), ch. 7. His discussion is based on an earlier and shorter presentation of Sandel's argument: Michael J. Sandel, 'The Case Against Perfection: What's Wrong with Designer Children, Bionic Athletes, and Genetic Engineering', *Atlantic Monthly* 292.3 (2004), reprinted in Savulescu and Bostrom, *Human Enhancement*, pp. 71–89.

[46] Harris, *Enhancing Evolution*, pp. 116–17.

[47] Ibid., p. 46.

Ageing is one of his examples: '[w]e do not die of old age but of the diseases of old age'.[48] If we systematically treated those diseases (therapy), the effect would be to extend the human lifespan (enhancement).

Harris' claim that the therapy-enhancement distinction is *morally* insignificant relates to his main ground for supporting enhancement, which informs a second criticism. He rejects out of hand Sandel's call to appreciate the giftedness of life and be open to the unbidden, and the connections between these attitudes and the virtues of humility, responsibility and solidarity. Where Sandel fears that the drive to mastery will place a limitless burden of responsibility on us, Harris regards this burden as one that autonomous beings with the ability to understand and manipulate the world cannot avoid. He considers Sandel's position a refusal of that proper responsibility.[49] His own core ethical criterion for its exercise is consequentialist: 'the overwhelming moral imperative for both therapy and enhancement is to prevent harm and confer benefit'.[50] By this criterion, enhancement is a 'moral duty'.[51]

Harris' third complaint against Sandel is that the latter's arguments will be used to prohibit enhancement technologies. 'Sandel can have the unbidden and welcome, but on condition he will let me and others have access to the bidden.'[52] In other words, alongside his consequentialist evaluation of human enhancement is a politically libertarian assertion of individual citizens' freedom to choose enhancements for themselves and their children, building on a defence of libertarianism earlier in the book.[53]

In various articles, Harris has applied this reasoning specifically to pharmacological cognitive enhancements, arguing that if enhanced cognitive functions are goods we think it right to pursue through education and training, it is inconsistent to object to their pursuit by means of drugs. He rather breezily dismisses concerns that their use would be unfair, unjust or discriminatory or could result in pressure or coercion to use them in

48 Ibid., p. 45.
49 Ibid., pp. 117–19.
50 Ibid., p. 58.
51 Ibid., ch. 2.
52 Ibid., p. 122.
53 Ibid., ch. 5. To my mind that defence of libertarianism is unpersuasive: it includes some assertion, a conflation of political liberty with libertarianism, and a rather unconvincing attempt to shift the burden of proof onto his opponents.

professional or educational settings.[54] Others have broadened Harris' benefit-based argument, arguing that cognitive and other neural enhancements could benefit not only the individuals who use them but also their communities and societies. For example, they could improve individuals' productivity and contribution to society,[55] or improve the quality of political life and democratic participation by enhancing the civic virtues of citizens.[56] States may therefore have an interest in promoting the use of such enhancements as a public good.[57]

The dispute between Sandel and many of his critics (including Harris) is really a clash of contrasting moral visions. Against Harris' consequentialism in which 'the overwhelming moral imperative … is to prevent harm and confer benefit', and his libertarian version of the right to autonomous choice,[58] Sandel explicitly maintains that 'the moral stakes in the enhancement debate are not fully captured by the familiar categories of autonomy and rights, on the one hand, and the calculation of costs and benefits, on the other'.[59] This dispute is unlikely to be easy to resolve conclusively,[60] though the theological analysis offered later in this chapter will in effect take sides, since the theological perspective I shall set out has more in common with Sandel's moral vision than Harris'.

This however highlights the second kind of objection anticipated by Sandel: that his argument depends on religious or theological presuppositions.[61] Although Sandel remarks that his concerns 'verge on theology' and acknowledges his debt to theologians like William F. May,[62]

[54] For example, John Harris, 'Is It Acceptable for People to Take Methylphenidate to Enhance Performance? Yes', *British Medical Journal* 338 (2009), pp. 1532–33; Harris, 'Chemical Cognitive Enhancement: Is It Unfair, Unjust, Discriminatory, or Cheating for Healthy Adults to Use Smart Drugs?' in Illes and Sahakian (eds.), *The Oxford Handbook of Neuroethics*, pp. 265–72.

[55] Allen Buchanan, 'Enhancement and the Ethics of Development', *Kennedy Institute of Ethics Journal* 18.1 (2008), pp. 1–34.

[56] Will Jefferson et al., 'Enhancement and Civic Virtue', *Social Theory and Practice* 40.3 (2014), pp. 499–527.

[57] Buchanan, 'Enhancement and the Ethics of Development'; Jefferson et al., 'Enhancement and Civic Virtue'; Nick Bostrom and Rebecca Roache, 'Smart Policy: Cognitive Enhancement and the Public Interest', in Savulescu et al. (eds.), *Enhancing Human Capacities*, pp. 138–49.

[58] Harris, *Enhancing Evolution*, ch. 5.

[59] Sandel, *The Case against Perfection*, p. 96. As this quotation shows, a similar point about a clash of moral visions could be made regarding non-consequentialist critiques based on autonomy and rights. Perhaps here we are led back to the familiar battle lines over Alasdair MacIntyre's critique of the 'Enlightenment project' in *After Virtue: A Study in Moral Theory* (London: Duckworth, 2nd edn, 1985).

[60] Though for an irenic attempt to find common ground between proponents and critics of enhancement, see Erik Parens, 'Toward a More Fruitful Debate about Enhancement', in Savulescu and Bostrom (eds.), *Human Enhancement*, pp. 181–97.

[61] This paragraph is a modified version of Messer, *Respecting Life*, p. 93.

[62] Sandel, *The Case against Perfection*, pp. 9, 45–50.

he does not believe that religious or theological commitment is needed to support his conclusions. Many people, he argues, use the language of 'sanctity' and 'gift' in respect of human life and abilities without attributing them to a divine Giver who sanctifies them. When we speak of an athlete's or musician's gift, for instance, we mean simply that 'whether he has nature, fortune or God to thank for it, [his] talent is an endowment that exceeds his control'.[63]

Sandel receives support on this point from Michael Hauskeller, who argues that it makes sense to regard human life and abilities as gifts for which we can feel gratitude, even without believing in a divine Giver, because '[g]ratitude is *not* primarily an emotion that is directed towards a particular person'.[64] Indeed, Hauskeller goes so far as to claim that 'non-personal gratitude is much more common, and much more *basic*, than interpersonal gratitude … [gratitude to particular people] is not where we have learned to feel grateful or how to feel grateful, nor what gratitude really means'.[65] This is not self-evident, however, and Hauskeller does not offer particularly strong reasons for believing it, as Guy Kahane observes: 'when a great good unexpectedly lands in our hands, this can occasion a distinctive kind of gladness; we can call such gladness "gratitude" if we wish, but this is more likely to mislead than to illuminate'.[66] According to Kahane, attitudes and practices such as openness to the unbidden and gratitude for the giftedness of life are not 'theism-neutral'.[67] However, he also argues that they are not unqualifiedly supported by theism either, because in a universe created and governed by God, nothing is unbidden in an absolute sense – at most, it is only unbidden relative to us. Only in a Godless universe can events be absolutely unbidden, and then it makes no sense to be grateful for them.[68]

I agree with Kahane that it is hard to see why this kind of absolute unbiddenness in a Godless universe should be welcomed, though as Hauskeller suggests, even in a Godless universe there could be reasons why the drive for

[63] Ibid., p. 93.

[64] Hauskeller, 'Human Enhancement', p. 67, emphasis original.

[65] Ibid., p. 69, emphasis original.

[66] Guy Kahane, 'Mastery without Mystery: Why There Is No Promethean Sin in Enhancement', *Journal of Applied Philosophy* 28.4 (2011), pp. 355–68 (366 n. 41).

[67] Ibid., passim.

[68] Ibid., pp. 357–61.

mastery criticized by Sandel should be regarded with suspicion.[69] Be that as it may, in a Christian theological perspective, the kinds of openness to the unbidden and gratitude for the giftedness of life that Sandel advocates certainly do not require what Kahane calls the 'absolutely unbidden'. They are perfectly consistent with the recognition that not everything in our lives or our world is under *our* control, but is (albeit in complex ways, which may be hard for us to discern) in the hands of a good and loving God.

The attitudes of openness to the unbidden and gratitude for the giftedness of life, then, are consistent with a Christian theological vision (*pace* Kahane), but it is more doubtful whether they are, as Sandel believes, free from implicit dependence on such a vision.[70] As we have seen, he offers a range of ways to understand concepts like 'gift', encompassing various different and perhaps incompatible views of life and nature, which are likely to have different practical outworking. It is not clear that the specific things Sandel wishes to say about life can simply be hitched up, willy-nilly, to just any of these understandings. They certainly seem closest to the more specifically theological ones, as is signalled by his use of the word 'talent' in articulating an allegedly non-theological notion of 'gift'. While 'talent' in everyday speech often means no more than 'ability', it has more specific origins in a parable of Jesus in which the 'talents' were sums of money entrusted to servants by a ruler, to whom the servants were accountable for the use of what they held in trust (Mt. 25:14-30). Perhaps, then when the language of 'gift' or 'talent' has the kind of resonance that Sandel wishes to ascribe to it, it is because of its origins in a specifically Christian narrative. Maybe the language of gift is not so easy to unhitch from its theological moorings as he thinks.

Theological evaluation

The ethical debates surveyed in the previous section raise three core questions. First, is there a coherent and morally significant distinction between therapy and enhancement? Second, how should we ethically evaluate therapies and

[69] Hauskeller, 'Human Enhancement', pp. 71–76, cites Voltaire's maxim that 'the better is the enemy of the good' to argue persuasively that a constant striving to improve on human nature could breed a chronic and corrosive kind of dissatisfaction with even the best things we presently experience, and inhibit us from feeling at home in the world. There are suggestive links to be made here with Dietrich Bonhoeffer's theological critique of 'radicalism', explored further below.

[70] This paragraph is taken, with modifications, from Messer, *Respecting Life*, pp. 93–94.

enhancements – by what criteria, guided by what kind of moral vision? Third, what normative conclusions will such an ethical evaluation yield: what should we conclude about the rights and wrongs of therapeutic interventions in the brain and neural enhancements? The remainder of the chapter is an attempt to answer these three questions within the frame of the Christian theological tradition articulated throughout this book.

Therapy and enhancement

A standard definition of enhancements is given by Eric Juengst: 'interventions designed to improve human form or functioning beyond what is necessary to sustain or restore good health.'[71] By implication, therapies can presumably be understood as interventions designed to sustain or restore good health.[72] If so, then as Sandel and Harris recognize, what we say about the therapy-enhancement distinction depends on how we understand health and disease.[73] They both link the distinction to an understanding of health as normal or species-typical functioning,[74] which Harris traces to Norman Daniels and Christopher Boorse.[75] He is highly critical of this way of conceptualizing health, claiming among other things that it entails the view that the diseases of old age (being species-typical) are not really diseases and treating them would not constitute therapy.[76] Boorse, though, would have little difficulty answering

[71] Eric Juengst, 'What Does Enhancement Mean?' in Erik Parens (ed.), *Enhancing Human Traits: Ethical and Social Implications* (Washington, DC: Georgetown University Press, 1998), pp. 29–47 (29).

[72] Of course, therapies and enhancements are not the only possible kinds of intervention, but they are the two that matter for our purposes. More to the point, it might be controversial to include the *sustaining* of good health in the definition of therapy. Some authors, such as Harris (*Enhancing Evolution*, p. 21) define interventions designed to sustain good health (such as vaccinations) as enhancements, because it is normal for humans to get infectious diseases and those who are given vaccinations are not usually ill when they receive them. But this seems an implausibly narrow understanding of the therapeutic, which gives the impression of being adopted at least partly for the purpose of discrediting the therapy-enhancement distinction. However, if someone insisted on reserving the term 'therapy' for interventions to treat already-existing diseases, it would be possible (though clumsy) to use a term like 'health-promoting interventions' to include both treatments and preventative measures, without affecting the substance of the following argument.

[73] On the philosophical accounts discussed in the following paragraphs, see further Neil Messer, *Flourishing: Health, Disease and Bioethics in Theological Perspective* (Grand Rapids, MI: Eerdmans, 2013), ch. 1.

[74] Sandel, *The Case against Perfection*, p. 47; Harris, *Enhancing Evolution*, ch. 3.

[75] For example, Christopher Boorse, 'Health as a Theoretical Concept', *Philosophy of Science* 44.4 (1977), pp. 542–73; Norman Daniels, 'Normal Functioning and the Treatment-Enhancement Distinction', *Cambridge Quarterly of Healthcare Ethics* 9 (2000), pp. 309–22.

[76] Harris, *Enhancing Evolution*, p. 45.

Harris' critique, because he does not – as Harris seems to think – define just any statistically species-typical condition as health. (That would entail some even more obviously absurd conclusions than those Harris identifies.) Boorse is a little more careful than that: in his theory, health is the absence of disease, and diseases are internal states which interfere with 'species-typical *contribution[s] to survival and reproduction*, characteristic of the organism's age'.[77]

Harris' critique does, however, obliquely indicate one of the commonly recognized problems with Boorse's account. Boorse intends his definitions of health and disease to be objective, value-free and definable in purely scientific terms, but critics frequently charge that he cannot deliver on that aim:[78] part of what we mean by disease just is that (other things being equal) it is a bad condition to be in. Value, in short, cannot be excluded from the concepts of health and disease.[79] In which case, what values are relevant? Harris has a quick answer: the relevant value is our 'rational preference not to be harmed'.[80] This renders definitions of disease and therapy-enhancement distinctions unnecessary, he thinks, since we can move directly from this value 'to a rejection of harm and an acceptance of benefit whether called therapy or enhancement'.[81] It is unclear whether he thinks the meaning of the human good is exhausted by benefit and the absence of harm (a view that since Bentham's day has been widely criticized as impoverished) or whether he thinks there are other aspects of the human good, but only benefit and harm are relevant to health (in which case, it is not obvious why other aspects are irrelevant). In any event, even a cursory survey of the philosophical literature on health and disease reveals a range of richer accounts of the values at stake

[77] Christopher Boorse, 'What a Theory of Mental Health Should Be', *Journal for the Theory of Social Behavior* 6 (1976), pp. 61–84 (62, emphasis added).

[78] See, for example, K. W. M. Fulford, 'What Is (Mental) Disease? An Open Letter to Christopher Boorse', *Journal of Medical Ethics* 27 (2001), pp. 80–85.

[79] This difficulty with Boorse's account is acknowledged (at least up to a point) by Norman Daniels, who used Boorse's definitions in his earlier work but in more recent publications has turned instead to Jerome Wakefield's account, which combines factual and evaluative elements in defining diseases or disorders as 'harmful dysfunctions': Norman Daniels, *Just Health: Meeting Health Needs Fairly* (Cambridge: Cambridge University Press, 2008), pp. 29–78; Jerome C. Wakefield, 'The Concept of Mental Disorder: On the Boundary between Biological Facts and Social Values', *American Psychologist* 47.3 (1992), pp. 373–88. Daniels believes that the 'dysfunction' component of Wakefield's definition retains the element of scientific objectivity needed to ground his, Daniels', account of distributive justice in health care.

[80] Harris, *Enhancing Evolution*, p. 45.

[81] Ibid., p. 46.

in health. Harris would no doubt have little time for some, such as Christopher Megone's Aristotelian account.[82] But even by his own lights, he should take more seriously an account like that of Lennart Nordenfelt, in which health is defined in terms of welfare or individuals' ability to realize their 'vital goals'.[83]

In short, there is much more to be said about health, disease and the therapy-enhancement distinction than Harris gets around to saying. But how might we think Christianly about these things?[84] Theological accounts of health and disease are fewer and further between than philosophical treatments, which is surprising when one considers what rich resources for understanding health and disease are provided by the Christian tradition. One source of such understanding may be found in Christian practices of healing and care of the sick, a tradition of practice that can be traced back to the New Testament and the beginnings of the Christian Church. Another is the Christian doctrine of creation, in which human beings are understood as the good creatures of a loving God, created for the fulfilment of characteristic goods, goals and purposes – ultimately, life with God. The doctrine of creation therefore also supports a teleological view of human life, though in a rather different way from the Aristotelian teleology expounded by Megone.

Theological and practical resources such as these ground an account in which humans can be understood as *creatures* of a particular kind, created with characteristic ends, goals or purposes. Our good or flourishing consists in the fulfilment of our proper creaturely ends. We have an ultimate end, eternal life with God, but also various 'penultimate' purposes appropriate to the kind of creature we are.[85] In Karl Barth's terms, the command of God the Creator sets us free to live before God, to live in fellowship with one another, to live the kind of integrated 'psycho-physical' life characteristic of human creatures, and to live limited lives in particular places and times, with particular vocations.[86]

[82] Christopher Megone, 'Aristotle's Function Argument and the Concept of Mental Illness', *Philosophy, Psychiatry, and Psychology* 5.3 (1998), pp. 187–201; idem, 'Mental Illness, Human Function, and Values', *Philosophy, Psychiatry, and Psychology* 7.1 (2000), pp. 45–65.

[83] For example, Lennart Nordenfelt, *On the Nature of Health: An Action-Theoretic Approach* (Dordrecht: Kluwer Academic, 2nd edn, 1995).

[84] For the following theological account see Messer, *Flourishing*, chs. 3, 4.

[85] On 'ultimate' and 'penultimate', see Dietrich Bonhoeffer, *Ethics* (Ilse Tödt et al. (ed.); Reinhard Krauss, Charles C. West and Douglas W. Stott (trans.); Dietrich Bonhoeffer Works, 6; Minneapolis, MN: Fortress, 2005), pp. 146–70.

[86] Karl Barth, *Church Dogmatics* (Geoffrey W. Bromiley and Thomas F. Torrance (ed.); 13 vols, Edinburgh: T&T Clark, 1956–75), vol. 3.4.

In this picture, health will be understood as a penultimate, not ultimate, good, concerned with the fulfilment of some of our this-worldly goals and ends: in Barth's words, the 'strength for human life'.[87] Diseases and disorders can be understood as internal states or conditions which disrupt our physical or mental functions, so as to hinder or threaten the fulfilment of some penultimate goals of embodied creaturely life.[88]

A cautionary note must be sounded here, however. The theological perspective taken in this book will emphasize how severely both our creaturely finitude and our sinfulness limit our knowledge of God's good purposes for human life, except insofar as God makes those purposes known to us – centrally, in and through Jesus Christ. (This informs, among other things, the theological suspicion of ethics-as-human-project that I voiced in Chapter 3.) Moreover, we find it all too easy to misunderstand what God *has* made known to us. This suggests that a good measure of epistemological humility is needed when we make claims about human flourishing and health.

Among other things, this gives theological reasons for taking very seriously the critical questions raised by disability studies about concepts of health and disease. For example, some disability authors criticize the concept of 'normal function' (central to many accounts of health and disease, such as Boorse's) as an ideological construct which turns statistically typical features of human bodies or minds into norms by which the quality of everyone's life is judged. These norms, it is argued, work to the detriment of those whose bodies or minds differ from the majority: disadvantages that such individuals suffer are attributed to their supposedly dysfunctional bodies or minds while social, political or economic causes are conveniently ignored.[89] A further theological source for understanding health and disease, therefore, is the rich vein of theological writing on disability that takes up some of these critical perspectives,

[87] Ibid., p. 356.

[88] *Some* penultimate goals: this view of health and disease has a limited scope, and does not include every aspect even of penultimate human good. Therefore we should treat with considerable caution the tendency in much Christian literature on health and healing to speak of health as 'wholeness' or *shalom*, sometimes drawing on the World Health Organization (WHO)'s definition of health as 'a state of complete physical, mental, and social well-being': for example, John Wilkinson, *The Bible and Healing: A Medical and Theological Commentary* (Grand Rapids, MI: Eerdmans, 1998), pp. 11–13. The WHO definition is in World Health Organization, *Basic Documents* (Geneva: World Health Organization, 48th edn, 2014), p. 1, online at http://apps.who.int/gb/bd/ (accessed 18 April 2016).

[89] See, for example, Ron Amundson, 'Against Normal Function', *Studies in History and Philosophy of Biological and Biomedical Sciences* 31.1 (2000), pp. 33–53.

but in some respects extends and intensifies them.[90] To express such a critical perspective more concretely: if we describe a particular condition – say Down's syndrome or congenital deafness – as a disease or disorder, might that description reflect a prejudiced, overly narrow conception of the ways in which human creatures may flourish before God?

With that caveat, this theological account supports a therapy-enhancement distinction that can do real ethical work and make sense of the problems in view in this chapter. Consider again the various uses of stimulants like modafinil and methylphenidate. On the account of health and disease outlined above, it makes theological sense to describe narcolepsy as a condition that disrupts certain physical and mental functions and so hinders the living of an integrated, embodied human life: in Barth's language, a form of weakness opposed to the strength for life. If modafinil can be used to treat this condition or alleviate its harmful symptoms, it makes sense to call its use therapeutic. This can probably also be said of ADHD and its treatment with methylphenidate, though this is less clear-cut because of longstanding controversies about diagnostic criteria, possible overdiagnosis in some countries, and the possible overuse of methylphenidate as a treatment.[91] Be that as it may, the same can hardly be said about the inability to gain a first-class rather than upper second-class degree, and the use of drugs to raise the level of one's exam performance cannot therefore be described as a therapy.

An approach to theological evaluation

The second core question identified at the start of this section was how neural therapies and enhancements should be ethically evaluated: by what criteria,

[90] In particular, some theologians argue that the lives of people with cognitive impairments should challenge common assumptions about the value and purpose of human life more radically than one often finds in the non-theological disability literature. See, for example, Hans S. Reinders, *Receiving the Gift of Friendship: Profound Disability, Theological Anthropology, and Ethics* (Grand Rapids, MI: Eerdmans, 2008); Thomas E. Reynolds, *Vulnerable Communion: A Theology of Disability and Hospitality* (Grand Rapids, MI: Brazos, 2008); John Swinton, 'Known by God', in Hans Reinders (ed.), *The Paradox of Disability: Responses to Jean Vanier and the L'Arche Community from Theology and the Sciences* (Grand Rapids, MI: Eerdmans, 2010). There are obvious connections here with the questions about personhood and the value of human life that were raised in the context of disorders of consciousness in ch. 3, above.

[91] See Lawrence H. Diller, 'The Run on Ritalin: Attention Deficit Disorder and Stimulant Treatment in the 1990s', in Martha J. Farah (ed.), *Neuroethics: An Introduction with Readings* (Cambridge, MA: MIT Press, 2010), pp. 42–57; James M. Swanson et al., 'Attention Deficit Hyperactivity Disorder: Defining a Spectrum Disorder and Considering Neuroethical Implications', in Illes and Sahakian (eds.), *The Oxford Handbook of Neuroethics*, pp. 309–40.

guided by what kind of moral vision. In the theological approach taken in this book, the moral vision and criteria for such ethical evaluation will have their roots in the narrative at the heart of the Christian faith: the story of a creation brought into being through the sheer exuberant love of God and pronounced 'very good' by its Creator, yet whose goodness is obscured and flawed by the presence of evil in its various forms, including human sin; the story of God's saving love at work in the incarnation, death and resurrection of Christ to overcome the effects of sin and evil, and to inaugurate a new age that will be completed at the eschaton, in which humanity and all creation will reach their ultimate fulfilment. So one theological way to approach the ethical evaluation of technological practices such as neurotechnologies is to ask how they relate to that Christian narrative of creation, sin, salvation and future hope. Are they in line with the narrative: practices that celebrate and cherish the goodness of creation, could be offered in service to God as contributions to the healing of the world's brokenness, and could direct us towards God's promised good future? Are they, alternatively, opposed to the Christian narrative, directed instead to destructive or chaotic ends? Or are they attempted substitutes (however well-intentioned) for God's saving and perfecting work in creation? Do they purport to address what a theological vision will recognize as the problems of sin and evil from human resources alone, in ways that are not open to God's help? Do they offer alternative eschatologies in place of the future hope affirmed by Christian faith?

These questions, however, need some further specification if they are to help us with concrete ethical analysis. Elsewhere I have proposed the following more specific 'diagnostic questions' to assist this task of discerning how technological practices relate to the Christian narrative.[92]

1. *Is the practice a way of acting that is appropriate to creatures made in God's image (*imago dei*), or does it reflect a desire to be 'like God' (*sicut deus*) in the sense promised by the snake in the 'Fall' narrative (Gn. 3:5)?* This contrast of *imago dei* with *sicut deus* is drawn from Dietrich Bonhoeffer, who reads the snake's invitation to become like God as an incitement to the humans to refuse their status as creatures, and instead aspire to be gods.[93] In a sense,

[92] Messer, *Respecting Life*, pp. 37–42.

[93] Dietrich Bonhoeffer, *Creation and Fall: A Theological Exposition of Genesis 1–3* (Martin Rüter, Ilse Tödt and John W. de Gruchy (eds.); Douglas Stephen Bax (trans.); Dietrich Bonhoeffer Works, 4; Minneapolis, MN: Fortress, 2004), pp. 111–14. For further reflection on Bonhoeffer's theological critique of the 'knowledge of good and evil', see above, ch. 3.

it works: they really do become 'like God, knowing good and evil'; but it is a catastrophe, alienating them from God, each other, themselves and the created world. Human action that reflects the image of God will embody a recognition of our creaturely limits, yet also take seriously our calling and ability to act in the world, to make something of it. Human action that (explicitly or implicitly) refuses all human limits, that claims the power to make whatever we choose of ourselves and the world, looks more like an attempt to be 'like God'.

2. *What attitude does it display towards the material world (including our own bodies)?* This question is also prompted by Bonhoeffer, in particular the ultimate/penultimate distinction from his *Ethics*, briefly discussed already in Chapter 5.[94] For Bonhoeffer, it is vital to understand the relationship of the ultimate to the penultimate rightly, and he identifies two errors to be avoided. One he calls 'radicalism', which would sweep away the penultimate in the light of the ultimate, regarding this world as no longer of any importance. The opposite error is 'compromise', which is so taken up with life here and now that the ultimate disappears from view and ceases to have any practical bearing on the way we live and act in this world. Paying attention to Jesus Christ, incarnate, crucified and risen, shows both to be inadequate, and teaches us to hold the ultimate and penultimate in proper balance. Technological practices based on the assumption that no hope for human beings will be found anywhere except our material existence in this world – as for example when we use technologies in an attempt to preserve and prolong human lives at almost any cost – look very much like what Bonhoeffer means by 'compromise'. Other technological projects, though, seem motivated by a profound impatience with the limitations of embodied life, and a desire to escape from those limits as far as we possibly can. There is something akin to Bonhoeffer's 'radicalism' in such an attitude. To get the balance right will mean using technologies in ways that recognize the material world (including our bodies) as *good* and worth taking trouble over, but not imagining that human hopes are limited to what physical life in this age can offer.

[94] Bonhoeffer, *Ethics*, pp. 146–70.

3. *What attitude does it demonstrate towards our neighbours?* To use this as one of our diagnostic questions is to allow our analysis to be informed by one of the two great commandments identified by Jesus: 'You shall love your neighbour as yourself' (Mk 12:29-31 par., quoting Lev. 19:18).
4. *Is it good news to the poor?* Drawing attention to a central Christian vocation, this question directs the attention of our ethical analysis to the concerns of the economically poor, but also more generally to the most disadvantaged, vulnerable or marginalized among our neighbours.
5. *What attitude does it display towards past failures?* Many of the human problems we need to address are caused in one way or another by human wickedness, weakness, folly or error – by what Christians would recognize as aspects of sin. To imagine we can solve our problems in the future without an honest acknowledgement of the past failures that have contributed to those problems is to risk replicating those failures and their destructive consequences. To avoid such dangers, technological projects intended to address human problems should be undertaken in a spirit of honest acknowledgement of past failures, recognition that things must be different in the future, and openness to the help we need to ensure that they *are* different, wherever that help is to be found; in other words, a spirit that could be called repentant.

Evaluating therapeutic interventions in the brain

The account of health and disease outlined earlier suggests a clear theological mandate for therapeutic activity. The goods of embodied creaturely life are penultimate, but nonetheless real and important, goods. If diseases threaten the fulfilment of those goods, then there is an imperative to address those threats. In terms of my second diagnostic question, such activity can be seen as a way of appropriately valuing the goods of the material world (specifically, bodily life and health), as witnessed by the healing narratives of the New Testament and the long Christian tradition of healing ministry and care for the sick. The most obvious motivation for such therapeutic work is compassion. My third and fourth diagnostic questions affirm this motivation, connecting it with the Christian calling to love one's neighbour, in particular those neighbours rendered especially vulnerable or marginalized by their medical conditions.[95] As Sandel suggests, the therapeutic goals of such treatments set limits to their ambitions;[96] in the terms of the first diagnostic question, this offers some safeguard against their becoming attempts at godlike mastery. Therapeutic acts like the use of modafinil to treat narcolepsy, antidepressants or TBS to treat depression, or DBS to treat Parkinson's disease would therefore receive *prima facie* support from this theological analysis.

Christian support for therapy should not of course be unqualified or unconditional. For any particular treatment, familiar issues such as safety, efficacy, the balance of harms, costs and benefits, and informed consent must be addressed. These issues come into particularly sharp focus in relation to interventions in the brain, where scientific understanding is still very limited, but the suffering caused by dysfunctions can be particularly acute. In these circumstances, the attraction of heroically innovative experimental treatments may seem especially strong. The fifth diagnostic question, about attitudes to past failures, may then be a necessary reminder of the damage done by overconfident interventions (such as lobotomy) in the past. Asking this question reinforces the point made by the Nuffield Council on Bioethics: in the development of new neurotechnologies the real and much-needed virtue of inventiveness must be held in proper balance with the equally necessary virtues of humility and responsibility.[97] Furthermore, the fourth diagnostic question, about good news to the poor, keeps concerns about distributive justice and equity of access to treatments – particularly innovative and expensive ones – in view.

More generally, the first two diagnostic questions serve as reminders that therapeutic activity must be in keeping with the *penultimate* character of health as a human good. Alongside the Gospel healing narratives, the New Testament contains the story of Paul's thorn in the flesh, which was not taken from him despite his prayers and pleas – an experience which became an occasion for him to discover more fully God's power at work in him (2 Cor. 12:7-10).

[95] Note that the Christian meaning of loving one's neighbour is not exhausted by compassion, and could in some cases even be distorted by an *exclusive* focus on compassion: see, for example, Messer, *Respecting Life*, ch. 8. But on any reasonable account of neighbour-love, compassion will have a central place. This is discussed more fully in relation to one particular vulnerable and marginalized patient group – those with disorders of consciousness – in Chapter 5, above.

[96] Sandel, *The Case against Perfection*, pp. 46–49.

[97] Nuffield Council, *Novel Neurotechnologies*, pp. 84–89.

In similar vein, Karl Barth – who affirms in the strongest terms that diseases are disruptions of God's good creation and should be resisted – also maintains that they can remind us of our finitude and mortality, thereby pointing us towards an ultimate hope which lies beyond this mortal life.[98] Health is a real but penultimate good; if it is treated as an ultimate good to be pursued by any means necessary, Christians should recognize this as a kind of idolatry that may have destructive consequences for all concerned. Among other things, it follows that Christian faith should recognize the wisdom of judging that sometimes enough is enough: there may come a point in a patient's care when further life-prolonging interventions are futile or disproportionate. This theological perspective also makes it easier to say that even health goals which are good in themselves should not be pursued if the means used to achieve them are unjustifiable. For example, if a particular neural stem cell therapy makes use of human embryonic stem cells (entailing the destruction of human embryos), the question whether the means is unacceptable, even though the therapeutic goal is good and important, must at any rate be taken seriously.[99]

Evaluating neural enhancements

To illustrate how neural enhancements may be theologically evaluated using the same approach, there follows an extended analysis of one specific example, the use of pharmacological cognitive enhancement (PCE) by students (though I shall also refer to other uses of the same enhancement technologies). Some of the argument will be quite specific to that example, and indeed I shall suggest that the ethical assessment of enhancement technologies requires, not a blanket 'Yes' or 'No', but a prudential judgement sensitive to the specific features of each application. Therefore the analysis of other enhancement technologies and other uses might yield different results – though the theological approach followed here will also suggest a more general perspective on human enhancement as a whole.

[98] Barth, *Church Dogmatics*, vol. 3.4, pp. 366–74.

[99] This is one reason for Christians to welcome and support the development of induced pluripotent stem cell technology, a less ethically fraught technique. On these neural stem cell therapies, see Nuffield Council, *Novel Neurotechnologies*, pp. 36–40. On the theological ethics of human embryonic stem cell research, see Messer, *Respecting Life*, ch. 4.

The first of my diagnostic questions draws a contrast between acting as creatures made in God's image and trying to be 'like God'. Recall Sandel's chief complaint against attempts at human enhancement: that they display a Promethean drive to mastery,[100] a drive that could be expressed theologically as a desire to be *sicut deus*. This ambition is unabashed in some transhumanist literature,[101] and both philosophical and theological commentators have drawn attention to the religious and theological aspirations expressed in this literature.[102] It would be highly implausible to claim that this is also the conscious motivation of most people using enhancements in more mundane ways, such as students using PCE in the hope of gaining better grades. Yet if such relatively mundane practices are seen as part of a larger cultural trend towards human technological enhancement, the drive for mastery may nonetheless be implicit in them, as Sandel suggests. This by itself does not give a conclusive reason to reject a practice such as PCE, but it does give grounds for caution and suspicion.

The second diagnostic question is about the attitudes displayed by technological projects towards the material world, including our own bodies. Many enhancement projects seem to manifest a general impatience with the limits of embodied creaturely existence (to put it no more strongly than that). Like the aspiration to become *sicut deus*, this impatience with life here and now is explicit in the transhumanist literature, for example, when human nature is described as 'a work-in-progress, a half-baked beginning that we can learn to remold in desirable ways';[103] but is also evident in many arguments for more mundane enhancement practices such as PCE.[104] It may make perfect sense if one denies any normative significance to nature on the grounds that 'we are the contingent, unbidden products of natural selection, a process that is driven by reproductive

[100] Sandel, *The Case against Perfection*, pp. 26–27.

[101] See, for example, Nick Bostrom, 'Are We Living in a Computer Simulation?', *Philosophical Quarterly* 53.211 (2003), pp. 243–55; Ray Kurzweil, *The Singularity Is Near* (London: Duckworth, 2005); Harris, *Enhancing Evolution*, pp. 62, 71, speculating about a future in which enhanced immortal individuals coexist with mortals. Although Harris does not identify himself with a transhumanist agenda, he remarks that 'If the consequence of [enhancement] is that we become transhumans, there is nothing wrong with that' (ibid., p. 39).

[102] For example, James Hughes, 'Contradictions from the Enlightenment Roots of Transhumanism', *Journal of Medicine and Philosophy* 35 (2010), pp. 622–40; Brent Waters, *From Human to Posthuman: Christian Theology and Technology in a Postmodern World* (Farnham: Ashgate, 2006).

[103] Bostrom, 'Transhumanist Values', p. 4; see also Bostrom, 'Why I Want to Be a Posthuman when I Grow Up', in Bert Gordijn and Ruth Chadwick (eds.), *Medical Enhancement and Posthumanity* (Dordrecht: Springer, 2008), pp. 107–37.

[104] For example, Harris, 'Chemical Cognitive Enhancement'.

fitness, not by the good.'[105] But it will not do for Christians, who can certainly agree that biologically, humans are the contingent products of natural selection, but who must also affirm that our embodied existence in this world reflects (however imperfectly and ambiguously) the good purposes of a loving Creator. Theologically understood, this impatience with the limitations of embodied humanity expresses what Bonhoeffer calls a radicalism that 'arises from a conscious or unconscious hatred of what exists' and 'cannot forgive God for having created what is.'[106]

Some nuance is required here, however, because a Christian doctrine of creation obviously does not claim that everything about embodied human life reflects God's good purposes. Christian faith regards the world and human life as good but flawed, in need of healing and transformation. Therefore, appropriate gratitude for the gift of our finite, embodied creaturely existence does not mean simply accepting every aspect of the way we are. This is why William F. May can emphasize the need for a proper *balance* between accepting and transforming love in parenting, or between 'beholding' and 'moulding' in our engagement with nature.[107] In discerning when acceptance or beholding is the proper response, and what we should instead seek to transform or mould, one helpful guideline – again – is the therapy-enhancement distinction. Theologically articulated, that distinction teaches us to discriminate between limitations that inhibit our kind of flourishing, and those that are simply aspects of our kind of creaturely existence. It may be quite right to be impatient with limitations of the first kind (which include diseases and disorders) and to use human understanding and skill to try and overcome them. But to limits of the second kind, this would be an inappropriate response, reflecting the kind of impatience with creaturely existence as such that Bonhoeffer criticizes. Moreover, there is a rich diversity of ways to flourish as creatures of our kind, which means that human people have many different gifts, and some have abilities that others lack. That too should be a cause for celebration rather than regret.

Considering the specific case of PCE in education, presumably students who use it aim in some sense to overcome limitations on their academic performance; but are those the kinds of limitation that really threaten or

[105] Kahane, 'Mastery without Mystery', p. 361; see also Harris, *Enhancing Evolution*, p. 35. As I noted earlier, however, even on such a view there could still be reasons for objecting to the drive for mastery: see above, n. 67.

[106] Bonhoeffer, *Ethics*, p. 155.

[107] Quoted by Sandel, *The Case against Perfection*, pp. 49–50.

inhibit one's flourishing as a human creature? Or, if students experience the limits to their academic performance as inhibiting their human flourishing, might that instead be because of a social context that is in some way distorted or oppressive? This line of thought leads on to our third question.

What attitude does PCE demonstrate towards our neighbours? We might gain some purchase on this question by asking what motivates students to take smart drugs. In the British and Irish survey cited earlier, the most commonly reported motivations were to 'enhance cognition', 'offset sleep deprivation' and 'enhance mood' (though curiosity was also cited as a motivation).[108] Although the study found substantial 'resilience' to PCE (that is, low levels of actual use despite students' awareness of PCE and interest in it), the authors doubt that this reflects lack of interest in cognitive enhancement as such, since far more students used caffeine tablets or energy drinks for these purposes. They suggest that low levels of use may reflect lack of availability and perhaps concerns about side effects or illegality. They also conjecture that

> PCE use patterns might look very different if students found PCE to make a remarkable difference in their academic performance and achievement. We would expect to find a higher rate of consistent PCE use under such circumstances; more demand for, and wider availability of PCEs in universities; and a higher likelihood of conversion from interest to use.[109]

This conjecture finds some support from the voices of British students who were using PCE (and apparently did believe it made a difference to their performance), as quoted in a recent media report.[110] These voices convey a powerful sense of the struggle to cope with a high-pressure academic environment, and the need to gain as good a degree result as possible in order to survive in a competitive job market.[111] It should be emphasized that these are anecdotal reports rather than data from systematically designed qualitative research.[112] Nonetheless, there is a certain resonance between these voices and

[108] Singh et al., 'Robust Resilience and Substantial Interest'.

[109] Ibid., p. 10.

[110] Cadwalladr, 'Students Used to Take Drugs to Get High'.

[111] For example, 'Suzy': 'I shouldn't even be here. I didn't even want to go to university but everyone said I should. And the work! It's just… there's so much of it! I feel like I wouldn't even have a chance if it wasn't for modafinil.' Or 'Daniel': 'My dad is really successful. And he got to where he is today with a 2:2 … You wouldn't get a look in the door with most jobs with that these days.' Both quoted by Cadwalladr, 'Students Used to Take Drugs to Get High'.

[112] As Singh et al. observe ('Robust Resilience and Substantial Interest', pp. 1–2), there is a shortage of research evidence on PCE use outside the USA, which can make public and policy debates over-reliant on media reporting and anecdotal evidence. They emphasize the need for more research.

contemporary British public discourse about higher education, which tends to identify its value chiefly in contributing to the economy, equipping graduates to be successful in a competitive job market, and therefore also promoting social mobility.[113] It might prove illuminating to set students' interest in PCE in the broader context of what that public discourse encourages them to think their university education is for.

Theological reflection on the purpose of university education will not of course deny the value of contributing to the economy, equipping graduates for employment or promoting social mobility. Indeed, as Stephen Heap observes, Christian concern for the common good, and for the poor and marginalized in particular, will lead theologians to affirm these as genuine goods. However, he goes on to remark that theologians will regard a near-exclusive emphasis on these goods as 'a very thin and … individualistic view of what universities are for'.[114] They will, rather, wish to locate goods such as contributing to the economy and promoting social mobility in a broader and richer picture of the value and purpose of university education.[115] Various theologians have argued that an essential part of that picture is the cultivation of the wisdom and other intellectual, political and social virtues that will equip students to

[113] As Stephen Heap observes, these are the dominant (though not completely exclusive) emphases in recent policy documents concerning higher education in England: in particular, Lord Browne of Madingley (Chair), *Securing a Sustainable Future for Higher Education: An Independent Review of Higher Education Funding and Student Finance* (October 2010), online at https://www.gov.uk/government/publications/the-browne-report-higher-education-funding-and-student-finance, Department for Business, Innovation and Skills, *Higher Education: Students at the Heart of the System* (June 2011), online at https://www.gov.uk/government/publications/higher-education-students-at-the-heart-of-the-system-2 and Department for Business, Innovation and Skills, *Fulfilling Our Potential: Teaching Excellence, Social Mobility and Student Choice* (November 2015) https://www.gov.uk/government/consultations/higher-education-teaching-excellence-social-mobility-and-student-choice (all accessed 22 April 2016). For administrative reasons, the policy documents are limited to England, but the discourse they reflect crosses national borders within (and indeed beyond) the UK. See Stephen Heap, 'New Directions for Universities?', *Theology* 119.4 (2016), pp. 244–52.

[114] Heap, 'New Directions for Universities?', p. 248.

[115] There is a rich vein of recent theological literature on this topic, including (e.g.): Margaret Atkins, 'For Gain, for Curiosity or for Edification: Why Do We Teach and Learn?', *Studies in Christian Ethics* 17.1 (2004), pp. 104–17; Gavin D'Costa, *Theology in the Public Square: Church, Academy and Nation* (Oxford: Blackwell, 2005); David Ford, *Christian Wisdom: Desiring God and Learning in Love* (Cambridge: Cambridge University Press, 2007); Mike Higton, *A Theology of Higher Education* (Oxford: Oxford University Press, 2012); Rowan Williams, 'What Is a University?' (lecture given at Wuhan University, China, 13 October 2006), online at http://rowanwilliams.archbishopofcanterbury.org/articles.php/1468/china-universities-have-essential-role-in-public-life (accessed 21 April 2016). The portion of this literature produced by Anglican theologians is helpfully surveyed by Heap, 'New Directions for Universities?'. As he emphasizes, this broader vision of higher education is not the exclusive preserve of theologians; accounts that are similar in some respects can be found in the work of others such as Stefan Collini, *What Are Universities For?* (London: Penguin, 2012).

be good citizens. This is one way in which universities can contribute to the common good, but there are others, such as '[using] their disciplines, learning and wisdom to imagine what good futures might look like and how they can be worked for'.[116]

In this broader and richer vision of the goods that higher education exists to foster, the use of PCE makes less sense than it does within the thinner, more individualistic and agonistic view common in contemporary public discourse. In this broader picture, PCE looks not so much wrong as beside the point: it is simply not directed to the most important goods that higher education exists to foster.[117] If so, that in itself will be a reason to resist its use, because were it to become an established practice, it could divert effort and attention from practices that would foster the more important goods. If you have a technology that promises a short-cut to better grades, it is tempting to see better grades as the chief point of the exercise. To someone with a hammer, everything looks like a nail.[118]

My fourth diagnostic question asks whether PCE would be good news to the poor, and by extension to those particularly vulnerable or marginalized in other ways. This question draws attention to concerns about justice and equity, which feature prominently in debates about cognitive enhancement.[119] If it became yet another opportunity to which the wealthy or well-connected had privileged access (as they already do to good education and the advantages it brings), it could increase social inequality and prove distinctly *bad* news for the poor.

[116] Heap, 'New Directions for Universities?', p. 247, citing David Ford, 'Knowledge, Meaning and the World's Great Challenges: Reinventing Cambridge University in the Twenty-First Century' (The Gomes Lecture, 2003), online at http://www.ideaofauniversity.com/resources/ (accessed 22 April 2016).

[117] Admittedly (as we saw earlier), authors such as Jefferson et al. ('Enhancement and Civic Virtue') have proposed that PCE could actually *contribute* to wider civic goods, by enabling citizens to be more intelligent and better informed in their democratic participation. But these proposals are highly speculative, far removed from what we know of the present realities of PCE use, and it is not clear that the factors inhibiting the development of civic virtue really are those that could plausibly be remedied by technological enhancements.

[118] This analysis has something in common with Alasdair MacIntyre's distinction between goods internal and external to a practice: *After Virtue*, pp. 186–91. However, MacIntyre's distinction draws the line too sharply for the point I am making. The goods that I have suggested are over-emphasized in a context where PCE use is commonplace are not (completely) external to the practice of education; they are simply not the most central or important goods that education exists to realize.

[119] See, for example, Harris, 'Chemical Cognitive Enhancement', which focuses on these and related concerns, adopting a rather dismissive attitude towards objections based on them.

Yet some advocates of enhancement technologies argue that their use might actually reduce injustice and inequality.[120] For one thing, as noted earlier, there is some evidence that PCEs have more impact on individuals whose baseline cognitive abilities are lower, in which case they might mitigate the effect of inequalities in ability.[121] Moreover, Anders Sandberg suggests that technological cognitive enhancements could prove quicker and cheaper routes to the benefits that '"conventional" means of enhancing cognition', such as education and training, achieve in more costly and time-consuming ways.[122] If so, they could serve to reduce rather than increase social inequality.[123] Those who take the broader and richer view of universities outlined above are likely to find Sandberg's description of education as a costly and time-consuming cognitive enhancement excessively reductive. But leaving that aside, it seems that questions of distribution and access to enhancement technologies are contingent matters, and focusing on them will not (by itself) resolve the ethical questions about PCE.

My final diagnostic question was about attitudes to past failures: I suggested that technological projects ought to be undertaken in a spirit that could be called repentant. It is worth emphasizing that in theological perspective the practice of repentance presupposes the promise of forgiveness and the possibility of a new beginning, by God's grace. Therefore, to act in this way is fundamentally to act in a spirit of hope. What I described as a repentant spirit will take human sinfulness, fallibility and frailty, and the failures that result from them, with full seriousness, but will refuse to regard any human failure as an ultimate catastrophe or a cause for despair.

Failure looms large in a rather different way in media reporting of PCE use, in which a powerful sense of the seriousness of academic failure and an overwhelming fear of it can be discerned in some student voices quoted.[124] It may not be too far-fetched to see in these fears a relocation of ultimate concerns onto this-worldly activities like education and employment. In a culture where

[120] For example, Anders Sandberg and Julian Savulescu, 'The Social and Economic Impacts of Cognitive Enhancement', in Savulescu et al. (eds.), *Enhancing Human Capacities*, pp. 92–112.

[121] See Sandberg and Savulescu, 'The Social and Economic Impacts', p. 95.

[122] Anders Sandberg, 'Cognition Enhancement: Upgrading the Brain', in Savulescu et al. (eds.), *Enhancing Human Capacities*, pp. 71–91 (72).

[123] Ibid., pp. 84–85.

[124] For example, the comments of 'Suzy', reported by Cadwalladr, 'Students Used to Take Drugs to Get High': see above, n. 109.

a transcendent horizon has largely disappeared from view – in which any form of hope extending beyond this world plays little part in directing most people's living and acting in the world – it is easy for this-worldly activities such as education, employment, relationships and family life to assume ultimate significance. The fulfilment of our human lives is judged entirely by our success or failure at these things. Christians should recognize this as a confusion of the ultimate and penultimate – in other words, a form of idolatry – with obviously destructive results. In Christian perspective, education, employment, relationships and the rest are real but *penultimate* goods. Failures associated with them are not ultimate disasters. Our ultimate fulfilment does not depend on our efforts. Christians who know themselves to be forgiven sinners, saved and promised the hope of eternal life by God's grace, are thereby set free to bear witness to a different attitude to past and future failures in their life and work in this world. So once again, PCE may turn out not to be the problem in itself, but a symptom of it and an attempt to address it; the kind of attempt, however, more likely to reinforce and replicate the problem than to solve it.

It has become apparent that my diagnostic questions, if asked about student use of PCE, can lead into wide-ranging reflections about (for example) the purpose and value of higher education and contemporary attitudes to success and failure. As I have already suggested, this could indicate that PCE in education is not the heart of the issue, but should be seen as a symptom of deeper societal and cultural issues, and an attempt to respond to them. If so, moral condemnation of the practice itself – and certainly of those who engage in it – may be a misplaced response. My analysis suggests, however, that there are nonetheless good reasons to resist its use, because it is the kind of response that risks replicating and reinforcing the problems to which it is directed. Answers to the five diagnostic questions, taken together, emphatically do not encourage the use of PCE in education.

As I have emphasized, though, much of this analysis has been specific to that particular use. Indeed, my analysis has suggested that to evaluate an enhancement technology theologically and ethically, we must consider it in relation to the practice whose performance it is meant to enhance – be that education, sport, medicine or some other.[125] If an enhancement technology

[125] A similar point in relation to another area of practice, namely sport, is made by Michael R. Shafer, *Well Played: A Christian Theology of Sport and the Ethics of Doping* (Eugene, OR: Pickwick, 2015).

truly supported the goods internal to the practice in which it was used, that use of it could be legitimate (subject to familiar considerations such as safety, efficacy and distributive justice). I have argued that there are good reasons to doubt that this is the case for students, but it is possible (I put it no more strongly than that) that the conclusion would be different in relation to other practices.[126] So this theological analysis of the ethics of neural enhancements does not yield a blanket 'Yes' or No' answer. Rather, as Brent Waters has also argued in relation to a different enhancement technology,[127] what is required is a Christian exercise of practical wisdom to evaluate each specific use of the technology. The diagnostic questions I have used in this chapter are meant to support just such exercises of practical wisdom.

However, my theological analysis does suggest a more general *suspicion* of neural enhancements, because they can all too easily be motivated by a basic dissatisfaction with the opportunities and limits of human creaturely existence as such. The use of such enhancement technologies easily becomes an expression of that radicalism which, as Bonhoeffer put it, 'cannot forgive God for having created what is'.[128] I argued earlier that the theological perspective of this book generates a strong presumption in favour of therapeutic activity; but when we turn to enhancement technologies, the presumption will be the other way. And that presumption will be particularly strong in relation to the more ambitious goals identified with transhumanism. These so overtly express that basic dissatisfaction with creaturely existence, and sometimes an explicit ambition to become godlike beings, that it is hard to avoid seeing them as counterfeits of the salvation and future hope which in Christian perspective are to be sought and found in Jesus Christ.

[126] The use of PCE by surgeons performing lengthy procedures is sometimes cited as one situation where it would bring significant benefits, for example, by Barbara Sahakian, quoted in Cadwalladr, 'Students Used to Take Drugs to Get High'.

[127] Brent Waters, 'Christian Ethics and Human Germline Genetic Modification', *Christian Bioethics* 18.2 (2012), pp. 115–25.

[128] Bonhoeffer, *Ethics*, p. 155.

7

Conclusion: Beyond Mutual Neglect

In the Introduction, I described this book as an initial attempt to repair the mutual neglect of Christian ethics and neuroethics. One of my aims has been to contribute theologically to a range of important neuroethical debates. By doing so, I have also tried to demonstrate the importance, value and potential of this kind of theological engagement with neuroethics.

In realizing those aims, this book has only been a beginning – but it has been a beginning. In Chapter 2, a discussion of the cognitive neuroscience of religion enabled me to outline how a Reformed theological tradition can best engage with questions raised by neuroscience. I argued for a mode of engagement in which the agenda is set by theology, and scientific findings and insights are critically appropriated to contribute to a theologically shaped understanding of what it is to be human. The following four chapters gave extended examples of how this approach might work in practice, in relation to ethical questions in particular.

Chapter 3 engaged with neuroscientific studies of moral judgement. These studies, as we saw, remain controversial and contested. However, I argued that insofar as they prove scientifically robust, the critical perspectives offered by these studies can make a limited but nonetheless useful contribution to the theological critique of ethics as merely human project, along lines suggested by Bonhoeffer and Barth. The same critical perspectives could also have a useful role to play in the theological transformation and reconstruction of ethics. Thus, the theological approach proposed in Chapter 2 proved well able to deal critically and constructively with a body of neuroscientific work which might at first seem quite corrosive of Christian ethics.

Chapter 4 addressed another area that appears to have challenging implications for Christian ethics: neuroscientific studies of conscious will, and the questions they raise about freedom and responsibility. I argued that such studies need not drive us to a hard determinist position or the view that mental causation and conscious will are illusions. More than this, however, I proposed that perplexing questions about freedom, autonomy and responsibility might be both complicated and enriched by being reframed in the context of core theological themes like providence, sin and grace. A theological account rooted in the thought of Augustine, developed by the Reformers and articulated by contemporary authors such as McFadyen and Couenhoven, proved to have valuable resources to support this task of reframing.

In Chapter 5, practical ethical questions about patient care, raised by functional imaging research on disorders of consciousness (DoCs), led quickly into more theoretical and speculative discussions of consciousness and the self. Theological accounts of the soul are clearly relevant to these discussions. However, as we also saw in the chapter, engagement with neuroscience (among other things) has prompted various recent authors to revisit the question of what an authentically Christian understanding of the soul might look like. Returning to the practical questions, I was critical of recent ethical arguments about the treatment of patients with DoCs. However, I not only criticized those arguments in their own terms. I also argued that the terms in which these debates are commonly framed – particularly questions about the *personhood* of these patients – are themselves unsatisfactory. I proposed an alternative way for a theological ethic to frame these issues, by taking its bearings from biblical texts like the parable of the Good Samaritan (Lk. 10:29-37). Such a reframed theological approach, I argued, may challenge widely held views about the ways in which the value of human lives or the protection due to them might be diminished by disorders of consciousness.

Finally, Chapter 6 explored practical questions about attempts to modify the working of the brain technologically. I argued that in theological perspective it is both coherent and important to distinguish between *therapeutic* interventions and those intended to *enhance* brain functions beyond their 'normal' or healthy levels. I proposed a way to evaluate both therapeutic and enhancement projects theologically by assessing how they stand in relation to the central Christian narrative of creation, sin, salvation and future hope. This approach, I argued, supports a strong, though not indefeasible, presumption in favour of therapeutic interventions. It suggests a more sceptical stance towards enhancement projects, though not a blanket rejection of them all. I argued for an exercise of discernment, guided by a theologically informed practical wisdom, in relation to each such project, though there is little to be said theologically in favour of the more grandiose ambitions of transhumanism.

This book has been a beginning, then, in repairing the mutual neglect of theology and neuroethics – but only a beginning. There is much more to be said theologically about all the topics I have discussed. On neural enhancement (Chapter 6), for example, there is more to be done to examine the various kinds of enhancement projects proposed in the literature – not only cognitive, but also mood and moral enhancement, as well as the large-scale ambitions of transhumanism – to identify the hopes, fears and ambitions implicit in these projects and ask how they might be evaluated theologically. Another major aspect of enhancement technologies not discussed at all in the present book is their military applications; there is certainly scope for theological reflection on such applications in the context of the ethics of modern warfare.

Chapter 5, on disorders of consciousness, raised questions about what is meant by consciousness, identity and the self, and how these concepts might be understood theologically – particularly in relation to Christian understandings of the soul. In this connection, there is more to be said about theological understandings of the self and the soul, and how these might relate to discussions elsewhere in the book about freedom, responsibility and sin. But discussions of identity and selfhood might also take a rather different shape were we to consider other clinical conditions such as degenerative brain disorders. The discussion about DoCs is mostly concerned with situations where there has been a sudden, apparently total or near-total, loss of conscious awareness. Degenerative disorders, on the other hand, may involve the gradual and progressive loss of aspects of identity, self-understanding or personality. They might therefore pose different challenges in speaking of our identity in relation to God and one another.[1]

The discussion of determinism, freedom and responsibility (Chapter 4) touched on questions about autonomy raised by the experience and scientific study of addiction. That discussion, though, points to wider questions about how autonomy should be understood in the light of neuroscientific

[1] Cf. John Swinton, *Dementia: Living in the Memories of God* (Grand Rapids, MI: Eerdmans, 2012).

perspectives on freewill and conscious agency. These wider questions could have significant implications for the central role that autonomy plays in health care ethics, among other things. In Chapter 4, I also alluded to, but did not pursue, the implications of neuroscience for legal responsibility. Joshua Greene and Jonathan Cohen think neuroscience 'changes nothing and everything' for the law;[2] there is an argument to be had about what (if anything) it changes for *Christian* perspectives on the law and criminal justice.

Chapter 3, on neuroscientific studies of moral judgement, showed one way in which critical perspectives arising from these studies might be appropriated into a theological critique of ethics-as-human-project and a theological reconstruction of ethics. However, apart from the critical theological perspective developed in that chapter, there are some quite different ways in which other Christian ethical approaches might engage with these scientific studies. A Christian natural law theory or virtue ethic, for example, might offer very different forms of engagement. Therefore, there is also a need for a critical dialogue *between* these different theological perspectives and their approaches to engaging with neuroscience.

Then, there are neuroethical topics and areas that I have not even touched on in the present book. For example, there is growing interest in the use of neuroimaging for 'brain reading' applications such as lie detection.[3] Although start-up companies have been formed to offer fMRI-based lie detection services, many neuroscientists maintain that the technology is not yet close to being reliably usable in such real-world applications, and there is controversy about how far it ever will be.[4] A theological ethical analysis of this area would not only take an interest in the moral acceptability of attempting to read people's minds (should that ever become feasible); it would also offer a critical perspective on the social and cultural factors that make it tempting to overclaim for the potential of neuroscience, and the concrete dangers of such overclaiming in

2 Joshua Greene and Jonathan Cohen, 'For the Law, Neuroscience Changes Nothing and Everything', in Martha J. Farah (ed.), *Neuroethics: An Introduction with Readings* (Cambridge, MA: MIT Press, 2010), pp. 232–58.

3 John-Dylan Haynes, 'Brain Reading: Decoding Mental States from Brain Activity in Humans', in Judy Illes and Barbara J. Sahakian (eds.), *The Oxford Handbook of Neuroethics* (Oxford: Oxford University Press, 2011), pp. 3–13.

4 Paul Root Wolpe, 'Is My Mind Mine? Neuroethics and Brain Imaging', in Vardit Ravitsky, Autumn Fiester and Arthur L. Caplan (ed.), *The Penn Center Guide to Bioethics* (New York: Springer, 2009), online at http://search.credoreference.com/content/entry/sppbioeth/is_my_mind_mine_neuroethics_and_brain_imaging/0 (accessed 11 November 2016).

practical fields like criminal law. Another area not touched on in this book is that of global perspectives on neuroethics:[5] what do questions of distributive justice and equity of access to neurotechnologies, for example, look like when set in the context of global health inequalities? A third emerging area that cries out for a theological treatment is the intersection between neuroethics and animal ethics. This list of topics still awaiting a theological engagement could be multiplied almost endlessly.

I set out in this book to begin repairing the mutual neglect of theology and neuroethics. By surveying a series of substantive neuroethical topics, I have tried to show how important the issues they raise are. In some cases, indeed, they are matters of literally vital concern, such as the treatment of patients with DoCs. However, we have also seen that the ways these issues are discussed in the literature are sometimes deeply problematic from the perspective of Christian ethics, so there is an urgent need for careful, critical and rigorous theological analysis of them. I have also argued that neuroethicists who think they can safely ignore theological perspectives actually have very poor grounds for doing so. More positively, my theological analyses of substantive neuroethical issues should suggest that neuroethicists who are prepared to take theology seriously may find it illuminating and enriching to do so. In short, I hope that this book does something to persuade both Christian ethicists and neuroethicists to move from mutual neglect to serious and careful engagement with each other – engagement which could prove creative, rewarding and productive for both sides.

[5] See Illes and Sahakian, *The Oxford Handbook of Neuroethics*, chs. 48–51.

Bibliography

Airedale N.H.S. Trust v Bland (1993) A.C. 789 House of Lords.

Albrecht, Gary L. and Patrick J. Devlieger, 'The Disability Paradox: High Quality of Life against All Odds', *Social Science and Medicine* 48.8 (1999), pp. 977–88.

Allen, Diogenes, 'Persons in Philosophical and Biblical Perspectives', in Malcolm Jeeves (ed.), *From Cells to Souls – and Beyond: Changing Portraits of Human Nature* (Grand Rapids, MI: Eerdmans, 2004), pp. 165–78.

Amundson, Ron, 'Against Normal Function', *Studies in History and Philosophy of Biological and Biomedical Sciences* 31.1 (2000), pp. 33–53.

Anderson, Ray S., 'On Being Human: The Spiritual Saga of a Creaturely Soul', in Warren S. Brown, Nancey Murphy and H. Newton Moloney (eds.), *Whatever Happened to the Soul? Scientific and Theological Portraits of Human Nature* (Minneapolis, MN: Fortress, 1998), pp. 175–94.

Anomaly, Jonny, 'Nietzsche's Critique of Utilitarianism', *Journal of Nietzsche Studies* 29 (2005), pp. 1–15.

Atran, Scott, *In Gods We Trust: The Evolutionary Landscape of Religion* (Oxford: Oxford University Press, 2002).

Appelbaum, Paul S., Loren H. Roth and Charles Lidz, 'The Therapeutic Misconception: Informed Consent in Psychiatric Research', *International Journal of Law and Psychiatry* 5.3–4 (1982), pp. 319–29.

Aquinas, Thomas, *The Summa Theologiae of St. Thomas Aquinas* (Kevin Knight (ed.); Fathers of the English Dominican Province (trans.); 2016). Online at http://newadvent.org/summa/index.html

Association for Psychological Science, 'APS Announces Third Replication Project' (28 October 2014). Online at http://www.psychologicalscience.org/index.php/publications/observer/obsonline/aps-announces-third-replication-project.html

Atkins, Margaret, 'For Gain, for Curiosity or for Edification: Why Do We Teach and Learn?', *Studies in Christian Ethics* 17.1 (2004), pp. 104–17.

Augustine, *The City of God* (Philip Schaff (ed.); Marcus Dods (trans.); *Nicene and Post-Nicene Fathers*, 1–02; Edinburgh: T&T Clark, 1886). Online at http://www.ccel.org/ccel/schaff/npnf102.i.html

Augustine, *On the Spirit and the Letter*, in *Anti-Pelagian Writings* (Philip Schaff (ed.); Peter Holmes, Robert Ernest Wallis and Benjamin B. Warfield (trans.); *Nicene and Post-Nicene Fathers*, 1–05; Edinburgh: T&T Clark, 1887). Online at http://www.ccel.org/ccel/schaff/npnf105.xi.html

Baars, Bernard J., 'Global Workspace Theory of Consciousness: Toward a Cognitive Neuroscience of Human Experience', *Progress in Brain Research* 150 (2005), pp. 45–53.

Baertschi, Bernard, 'The Burden of Self-Consciousness', *American Journal of Bioethics* 8.9 (2008), pp. 33–34.

Baertschi, Bernard and Alexandre Mauron, 'Genetic Determinism, Neuronal Determinism, and Determinism *Tout Court*', in Judy Illes and Barbara J. Sahakian (eds.), *The Oxford Handbook of Neuroethics* (Oxford: Oxford University Press, 2011), pp. 151–60.

Banks, William P. and Eve A. Isham, 'Do We Really Know What We Are Doing? Implications of Reported Time of Decision for Theories of Volition', in Walter Sinnott-Armstrong and Lynn Nadel (eds.), *Conscious Will and Responsibility: A Tribute to Benjamin Libet* (Oxford: Oxford University Press, 2010), pp. 47–60.

Bargh, John A. et al., 'Automaticity in Social-Cognitive Processes', *Trends in Cognitive Sciences* 16.12 (2012), pp. 593–605.

Barker, Roger A. and Alasdair Coles, 'The Ethical Issues of Trials of Neural Grafting in Patients with Neurodegenerative Conditions', in Judy Illes and Barbara J. Sahakian (eds.), *The Oxford Handbook of Neuroethics* (Oxford: Oxford University Press, 2011), pp. 455–65.

Barrett, Justin L., 'Cognitive Science of Religion: What Is It and Why Is It?' *Religion Compass* 1.6 (2007), pp. 768–86.

Barrett, Justin L., 'Cognitive Science of Religion: Looking Back, Looking Forward', *Journal for the Scientific Study of Religion* 50.2 (2011), pp. 229–39.

Barth, Karl, *Church Dogmatics* (G. W. Bromiley and T. F. Torrance (ed.), 13 vols., Edinburgh: T&T Clark, 1956–75).

Barth, Karl, *The Word of God and the Word of Man* (Douglas Horton Smith (trans.); New York: Harper, 1957).

Barth, Karl, *Protestant Theology in the Nineteenth Century* (Brian Cozens and John Bowden (trans.); Grand Rapids, MI: Eerdmans, 2002).

Bauby, Jean-Dominique, *The Diving Bell and the Butterfly* (Jeremy Leggatt (trans.); London: Fourth Estate, 1997).

Baumeister, Roy F., et al., 'Ego Depletion: Is the Active Self a Limited Resource?' *Journal of Personality and Social Psychology* 74.5 (1998), pp. 1252–65.

Baumeister, Roy F., et al., 'The Strength Model of Self-Control', *Current Directions in Psychological Science* 16.6 (2007), pp. 351–55.

Baumeister, Roy F. and Kathleen D. Vohs, 'Misguided Effort with Elusive Implications', *Perspectives on Psychological Science* 11.4 (2016), pp. 574–75.

Beauregard, Mario and Denyse O'Leary, *The Spiritual Brain: A Neuroscientist's Case for the Existence of the Soul* (New York: HarperCollins, 2007).

Beauregard, Mario and Vincent Paquette, 'Neural Correlates of a Mystical Experience in Carmelite Nuns', *Neuroscience Letters* 405 (2006), pp. 186–90.

Beauregard, Mario and Vincent Paquette, 'EEG Activity in Carmelite Nuns during a Mystical Experience', *Neuroscience Letters* 444 (2008), pp. 1–4.

Berg, Sanchia, 'Permanent Vegetative State: A Family's Agony', *BBC News Magazine* (23 September 2016). Online at http://www.bbc.co.uk/news/magazine-37444379

Berker, Selim, 'The Normative Insignificance of Neuroscience', *Philosophy and Public Affairs* 37.4 (2009), pp. 293–329.

Berridge, Kent C. and Terry E. Robinson, 'What Is the Role of Dopamine in Reward: Hedonic Impact, Reward Learning, or Incentive Salience?' *Brain Research Reviews* 28.3 (1998), pp. 309–69.

Biggar, Nigel, *The Hastening that Waits: Karl Barth's Ethics* (Oxford: Clarendon Press, 1993).

Biggar, Nigel, *Behaving in Public: How to Do Christian Ethics* (Grand Rapids, MI: Eerdmans, 2011).

Bird, Stephanie J., 'Ethical Challenges in Research: Another Look', *American Journal of Bioethics* 8.9 (2008), pp. 15–17.

Bird, Stephanie J., Michael Kalichman and Dena Plemmons (eds.), Theme Issue, 'Multidisciplinary Perspectives on Neuroethics', *Science and Engineering Ethics* 18.3 (2012).

Bishop, Jeffrey P. and Elliott Louis Bedford (eds.), Theme Issue, 'Medically Assisted Nutrition and Hydration: The Vegetative State and Beyond', *Christian Bioethics* 17.2 (2011).

Block, Ned, 'On a Confusion about a Function of Consciousness', *Behavioral and Brain Sciences* 18.2 (1995), pp. 227–47.

Block, Ned, *Consciousness, Function and Representation* (Cambridge, MA: MIT Press, 2007).

Block, Ned, 'What Was I Thinking?' *New York Times*, 28 November 2010.

Bloom, Paul, 'Religious Belief as an Evolutionary Accident', in Jeffrey Schloss and Michael J. Murray (eds.), *The Believing Primate: Scientific, Philosophical, and Theological Reflections on the Origins of Religion* (Oxford: Oxford University Press, 2009), pp. 118–27.

Boehm, Christopher, 'Explaining the Prosocial Side of Moral Communities', in Philip Clayton and Jeffrey Schloss (eds.), *Evolution and Ethics: Human Morality in Biological and Religious Perspective* (Grand Rapids, MI: Eerdmans, 2004), pp. 78–100.

Boethius, *Theological Tractates* (E. Capps, T. E. Page and W. H. D. Rouse (ed.); H. F. Stewart and E. K. Rand (trans.); Loeb Classical Library; London: William Heinemann, n.d.). Online at http://www.ccel.org/ccel/boethius/tracts

Boly, Melanie, Marcello Massimini and Giulio Tononi, 'Theoretical Approaches to the Diagnosis of Altered States of Consciousness', *Progress in Brain Research* 177 (2009), pp. 383–98.

Bonhoeffer, Dietrich, *Letters and Papers from Prison* (Eberhard Bethge (ed.); Reginald Fuller et al. (trans.); London: SCM, 1971).

Bonhoeffer, Dietrich, *Creation and Fall: A Theological Exposition of Genesis 1–3* (Martin Rüter, Ilse Tödt and John W. de Gruchy (ed.); Douglas Stephen Bax (trans.); Dietrich Bonhoeffer Works, 4; Minneapolis, MN: Fortress, 2004).

Bonhoeffer, Dietrich, *Ethics* (Ilse Tödt et al. (ed.); Reinhard Krauss, Charles C. West and Douglas W. Stott (trans.); Dietrich Bonhoeffer Works, 6; Minneapolis, MN: Fortress, 2005).

Boorse, Christopher, 'What a Theory of Mental Health Should Be', *Journal for the Theory of Social Behavior* 6 (1976), pp. 61–84.

Boorse, Christopher, 'Health as a Theoretical Concept', *Philosophy of Science* 44.4 (1977), pp. 542–73.

Bortolotti, Lisa, 'Review of *Self Comes to Mind* by Antonio Damasio', *Cognitive Neuropsychiatry* 17.2 (2012), pp. 191–95.

Bostrom, Nick, 'Are We Living in a Computer Simulation?' *Philosophical Quarterly* 53.211 (2003), pp. 243–55.

Bostrom, Nick, 'Transhumanist Values', *Journal of Philosophical Research* 30, suppl. (2005a), pp. 3–14.

Bostrom, Nick, 'The Fable of the Dragon-Tyrant', *Journal of Medical Ethics* 31.5 (2005b), pp. 273–77.

Bostrom, Nick, 'Why I Want to Be a Posthuman When I Grow Up', in Bert Gordijn and Ruth Chadwick (eds.), *Medical Enhancement and Posthumanity* (Dordrecht: Springer, 2008), pp. 107–37.

Bostrom, Nick and Rebecca Roache, 'Smart Policy: Cognitive Enhancement and the Public Interest', in Julian Savulescu, Ruud ter Meulen and Guy Kahane (eds.), *Enhancing Human Capacities* (Chichester: Wiley-Blackwell, 2011), pp. 138–49.

Boyer, Pascal, *Religion Explained: The Evolutionary Origins of Religious Thought* (New York: Basic Books, 2001).

Boyer, Pascal, 'Why Do Gods and Spirits Matter at All? Supernatural Gadgets and Social Mind Adaptations', in I. Pyysiaïnen and V. Anttonen (eds.), *Current Approaches in the Cognitive Science of Religion* (London: Continuum, 2002), pp. 68–92.

Bretherton, Luke, *Christianity and Contemporary Politics* (Chichester: Wiley-Blackwell, 2010).

Brown, Warren S., 'Neurobiological Embodiment of Spirituality and Soul', in Malcolm Jeeves (ed.), *From Cells to Souls – and Beyond: Changing Portraits of Human Nature* (Grand Rapids, MI: Eerdmans, 2004), pp. 58–76.

Brown, Warren S. and Brad D. Strawn, *The Physical Nature of Christian Life: Neuroscience, Psychology and the Church* (Cambridge: Cambridge University Press, 2012).

Brown, Warren S., Nancey Murphy and H. Newton Moloney (eds.), *Whatever Happened to the Soul? Scientific and Theological Portraits of Human Nature* (Minneapolis, MN: Fortress, 1998).

Browne of Madingley, Lord (Chair), *Securing a Sustainable Future for Higher Education: An Independent Review of Higher Education Funding and Student Finance* (October 2010). Online at https://www.gov.uk/government/publications/the-browne-report-higher-education-funding-and-student-finance

Buchanan, Allen, 'Enhancement and the Ethics of Development', *Kennedy Institute of Ethics Journal* 18.1 (2008), pp. 1–34.

Bulbulia, Joseph, 'Religiosity as Mental Time-Travel: Cognitive Adaptations for Religious Behavior', in Jeffrey Schloss and Michael J. Murray (eds.), *The Believing Primate: Scientific, Philosophical, and Theological Reflections on the Origins of Religion* (Oxford: Oxford University Press, 2009), pp. 44–75.

Buniak, Liana, Martina Darragh and James Giordano, 'A Four-Part Working Bibliography of Neuroethics: Part 1: Overview and Reviews – Defining and Describing the Field and Its Practices', *Philosophy, Ethics, and Humanities in Medicine* 9.9 (2014), DOI: 10.1186/1747-5341-9-9.

Bush, George, Presidential Proclamation 6158 (17 July 1990). Online at http://www.loc.gov/loc/brain/proclaim.html

Cadwalladr, Carole, 'Students Used to Take Drugs to Get High. Now They Take Them to Get Higher Grades', *The Observer* (15 February 2015). Online at http://www.theguardian.com/society/2015/feb/15/students-smart-drugs-higher-grades-adderall-modafinil

Carter Evan, C. and Michael E. McCullough, 'Publication Bias and the Limited Strength Model of Self-Control: Has the Evidence for Ego Depletion Been Overestimated?' *Frontiers in Psychology* 5.823 (2014), pp. 1–11.

Cartwright, Nancy, *The Dappled World: A Study of the Boundaries of Science* (Cambridge: Cambridge University Press, 1999).

Cavanaugh, William T., '"A Fire Strong Enough to Consume the House:" The Wars of Religion and the Rise of the Nation State', *Modern Theology* 11.4 (1995), pp. 397–420.

Chapman, Philip, 'Barth and Darwin: What Is Humanity?' *Theology and Science* 12.4 (2014), pp. 362–77.

Charland, Louis C., 'Affective Neuroscience and Addiction', *American Journal of Bioethics* 7.1 (2007), pp. 20–21.

Collini, Stefan, *What Are Universities For?* (London: Penguin, 2012).

Comte, Auguste, *System of Positive Polity* (Frederic Harrison (trans.); London: Longmans, Green and Co., 1875). Online at https://openlibrary.org/works/OL761013W/System_of_positive_polity

Congregation for the Doctrine of the Faith, 'Instruction on Respect for Human Life in Its Origin and on the Dignity of Procreation *Donum Vitae*' (22 February 1987).

Congregation for the Doctrine of the Faith, 'Responses to Certain Questions of the United States Conference of Catholic Bishops Concerning Artificial Nutrition and Hydration' (1 August 2007). Online at http://www.vatican.va/roman_curia/congregations/cfaith/documents/rc_con_cfaith_doc_20070801_risposte-usa_en.html

Cook, Christopher C. H., *Alcohol, Addiction and Christian Ethics* (Cambridge: Cambridge University Press, 2006).

Couenhoven, Jesse, 'The Necessities of Perfect Freedom', *International Journal of Systematic Theology* 14.4 (2012), pp. 396–419.

Couenhoven, Jesse, *Stricken by Sin, Cured by Christ: Agency, Necessity, and Culpability in Augustinian Theology* (Oxford: Oxford University Press, 2013).

Crick, Francis, *The Astonishing Hypothesis: The Scientific Search for the Soul* (New York: Charles Scribner's Sons, 1994).

Crick, Francis and Christof Koch, 'A Framework for Consciousness', *Nature Neuroscience* 6.2 (2003), pp. 119–26.

d'Aquili, Eugene G. and Andrew B. Newberg, 'The Neuropsychological Basis of Religions, or Why God Won't Go Away', *Zygon* 33.2 (1998), pp. 187–201.

d'Aquili, Eugene G. and Andrew B. Newberg, *The Mystical Mind: Probing the Biology of Religious Experience* (Minneapolis, MN: Fortress, 1999).

D'Costa, Gavin, *Theology in the Public Square: Church, Academy and Nation* (Oxford: Blackwell, 2005).

Damasio, Antonio, *Descartes' Error: Emotion, Reason and the Human Brain* (New York: Putnam, 1994).

Damasio, Antonio, *Self Comes to Mind: Constructing the Conscious Brain* (London: Vintage, 2012).

Daniels, Norman, 'Normal Functioning and the Treatment-Enhancement Distinction', *Cambridge Quarterly of Healthcare Ethics* 9 (2000), pp. 309–22.

Daniels, Norman, *Just Health: Meeting Health Needs Fairly* (Cambridge: Cambridge University Press, 2008).

Daniels, Norman, 'Reflective Equilibrium', in *The Stanford Encyclopedia of Philosophy* ed. Edward N. Zalta (Winter 2013). Online at http://plato.stanford.edu/archives/win2013/entries/reflective-equilibrium/

Darragh, Martina, Liana Buniak and James Giordano, 'A Four-Part Working Bibliography of Neuroethics: Part 2 – Neuroscientific Studies of Morality and Ethics', *Philosophy, Ethics, and Humanities in Medicine* 10.2 (2015), DOI: 10.1186/s13010-015-0022-0.

Darwin, Charles, *The Origin of Species* (1859; repr. New York: Gramercy, 1979).

Darwin, Charles, *The Descent of Man, and Selection in Relation to Sex* (2 vols; London: John Murray, 1871). Online at http://darwin-online.org.uk/content/frameset?pageseq=1&itemID=F937.1&viewtype=text

Davies, Oliver, *Theology of Transformation: Faith, Freedom and the Christian Act* (Oxford: Oxford University Press, 2013).

Dawkins, Richard, *The Extended Phenotype* (Oxford: Oxford University Press, 1982).

Dawkins, Richard, *The Selfish Gene* (Oxford: Oxford University Press, 2nd edn, 1989).

Dawkins, Richard, *The God Delusion* (New York: Houghton Mifflin, 2007).

de Jongh, Reinoud, et al., 'Botox for the Brain: Enhancement of Cognition, Mood and Pro-Social Behavior and Blunting of Unwanted Memories', *Neuroscience and Biobehavioral Reviews* 32 (2008), pp. 760–76.

de Waal, Frans, *Good Natured: The Origins of Right and Wrong in Humans and Other Animals* (Cambridge, MA: Harvard University Press, 1996).

de Waal, Frans, et al., *Primates and Philosophers: How Morality Evolved*, ed. Stephen Macedo and Josiah Ober (Princeton, NJ: Princeton University Press, 2006).

Deacon, Terrence W., 'Three Levels of Emergent Phenomena', in Nancey Murphy and William R. Stoeger (eds.), *Evolution and Emergence: Systems, Organisms, Persons* (Oxford: Oxford University Press, 2007), pp. 88–110.

Deane-Drummond, Celia, *Christ and Evolution: Wonder and Wisdom* (London: SCM, 2009).

Deane-Drummond, Celia, *The Wisdom of the Liminal: Evolution and Other Animals in Human Becoming* (Grand Rapids, MI: Eerdmans, 2014).

Deane-Drummond, Celia and Peter Manley Scott (eds.), *Future Perfect? God, Medicine and Human Identity* (London: T&T Clark, 2006).

Dehaene, Stanislas, et al., 'Conscious, Preconscious, and Subliminal Processing: A Testable Taxonomy', *Trends in Cognitive Sciences* 10.5 (2006), pp. 204–11.

DeHart, Paul J., *The Trial of the Witnesses: The Rise and Decline of Postliberal Theology* (Oxford: Blackwell, 2006).

Delkeskamp-Hayes, Corinna and Neil Messer (eds.), Theme Issue, 'Theological Anthropology and the Ethics of Human Germline Genetic Modification', *Christian Bioethics* 18.2 (2012).

Dennett, Daniel, *Consciousness Explained* (Boston, MA: Little, Brown and Co., 1991).

Dennett, Daniel, *Freedom Evolves* (London: Allen Lane, 2003).

Dennett, Daniel, *Breaking the Spell: Religion as a Natural Phenomenon* (New York: Viking, 2006).

Department for Business, Innovation and Skills, *Higher Education: Students at the Heart of the System* (June 2011). Online at https://www.gov.uk/government/publications/higher-education-students-at-the-heart-of-the-system–2

Department for Business, *Fulfilling Our Potential: Teaching Excellence, Social Mobility and Student Choice* (November 2015). Online at https://www.gov.uk/government/consultations/higher-education-teaching-excellence-social-mobility-and-student-choice

Di, H. B., et al., 'Cerebral Response to Patient's Own Name in the Vegetative and Minimally Conscious States', *Neurology* 68 (2007), pp. 895–99.

Dill, Brendan and Richard Holton, 'The Addict in Us All', *Frontiers in Psychiatry* 5.139 (2014), pp. 1–20.

Dillehay, Tom D., et al., 'Early Holocene Coca Chewing in Northern Peru', *Antiquity* 84.326 (2010), pp. 939–53.

Diller, Lawrence H., 'The Run on Ritalin: Attention Deficit Disorder and Stimulant Treatment in the 1990s', in Martha J. Farah (ed.), *Neuroethics: An Introduction with Readings* (Cambridge, MA: MIT Press, 2010), pp. 42–57.

Doctrine Commission of the Church of England, *The Mystery of Salvation* (London: Church House Publishing, 1997).

Donne, John, *Complete Poetical Works* (Delphi Classics, 2012).

Dresler, Martin, et al., 'Non-Pharmacological Cognitive Enhancement', *Neuropharmacology* 64 (2013), pp. 529–43.

Dunn, James D. G., *The Theology of Paul the Apostle* (Grand Rapids, MI: Eerdmans, 1998).

Eagleman, David, *The Brain: The Story of You* (Edinburgh: Canongate, 2015).

Eagleman, David, *The Brain with David Eagleman* (BBC4, 6 parts, 21 January–26 February 2016).

Ebert, Jeffrey P. and Daniel M. Wegner, 'Bending Time to One's Will', in Walter Sinnott-Armstrong and Lynn Nadel (eds.), *Conscious Will and Responsibility: A Tribute to Benjamin Libet* (Oxford: Oxford University Press, 2010), pp. 134–45.

Eklund, Anders, Thomas E. Nichols and Hans Knutsson, 'Cluster Failure: Why fMRI Inferences for Spatial Extent Have Inflated False-Positive Rates', *Proceedings of the National Academy of Sciences of the USA* 113.28 (2016), pp. 7900–5.

Eliot, George, *Middlemarch* (W. J. Harvey (ed.), London: Penguin, 1994).

Erasmus, 'On the Freedom of the Will', in *Luther and Erasmus* (E. Gordon Rupp and A. N. Marlow (trans.); Library of Christian Classics, 17; London: SCM, 1969), pp. 35–97.

Farah, Martha J., 'That Little Matter of Consciousness', *American Journal of Bioethics* 8.9 (2008), pp. 17–19.

Fernández-Espejo, Davinia and Adrian M. Owen, 'Detecting Awareness after Severe Brain Injury', *Nature Reviews Neuroscience* 14 (2013), pp. 801–9.

Fins, Joseph J., 'Clinical Pragmatism and the Care of Brain Damaged Patients: Toward a Palliative Neuroethics for Disorders of Consciousness', *Progress in Brain Research* 150 (2005), pp. 565–81.

Fins, Joseph J., 'Border Zones of Consciousness: Another Immigration Debate?' *American Journal of Bioethics* 7.1 (2007), pp. 51–54.

Fins, Joseph J., 'Neuroethics and Neuroimaging: Moving Toward Transparency', *American Journal of Bioethics* 8.9 (2008), pp. 46–52.

Fins, Joseph J., *Rights Come to Mind* (New York: Cambridge University Press, 2015).

Fins, Joseph J., et al., 'Neuroimaging and Disorders of Consciousness: Envisioning an Ethical Research Agenda', *American Journal of Bioethics* 8.9 (2008), pp. 3–12.

Fletcher, Joseph F., 'Four Indicators of Humanhood: The Enquiry Matures', *Hastings Center Report* 4.6 (1974), p. 4.

Frankfurt, Harry, 'Alternate Possibilities and Moral Responsibility', *Journal of Philosophy* 66 (1966), pp. 829–39.

Frankfurt, Harry, 'Freedom of the Will and the Concept of a Person', *Journal of Philosophy* 68.1 (1971), pp. 5–20.

Foddy, Bennett and Julian Savulescu, 'Addiction and Autonomy: Can Addicted People Consent to the Prescription of Their Drug of Addiction?' *Bioethics* 20.1 (2006a), pp. 1–15.

Foddy, Bennett and Julian Savulescu, 'Autonomy, Addiction and the Drive to Pleasure: Designing Drugs and our Biology: A Reply to Neil Levy', *Bioethics* 20.1 (2006b), pp. 21–23.

Foddy, Bennett and Julian Savulescu, 'Addiction Is Not an Affliction: Addictive Desires Are Merely Pleasure-Oriented Desires', *American Journal of Bioethics* 7.1 (2007), pp. 29–32.

Ford, David, 'Knowledge, Meaning and the World's Great Challenges: Reinventing Cambridge University in the Twenty-First Century' (The Gomes Lecture, 2003). Online at http://www.ideaofauniversity.com/resources/

Ford, David, *Christian Wisdom: Desiring God and Learning in Love* (Cambridge: Cambridge University Press, 2007).

Fulford, K. W. M., 'What Is (Mental) Disease? An Open Letter to Christopher Boorse', *Journal of Medical Ethics* 27 (2001), pp. 80–85.

Gazzaniga, Michael S., Richard B. Ivry and George R. Mangun, *Cognitive Neuroscience: The Biology of the Mind* (New York: W. W. Norton, 4th edn, 2012).

Geertz, Armin W., 'How *Not* to Do the Cognitive Science of Religion Today', *Method and Theory in the Study of Religion* 20 (2008), pp. 7–21.

Geertz, Armin W., 'When Cognitive Scientists Become Religious, Science Is in Trouble: On Neurotheology from a Philosophy of Science Perspective', *Religion* 39.4 (2009), pp. 319–24.

Glannon, Walter, *Brain, Body and Mind: Neuroethics with a Human Face* (Oxford: Oxford University Press, 2011).

Goertzel, Ben and Matthew Iklé (eds.), 'Special Issue on Mind Uploading', *Journal of Machine Consciousness* 4.1 (2012).

Goetz, Stewart, 'Substance Dualism', in Joel B. Green and Stuart L. Palmer (eds.), *In Search of the Soul: Four Views of the Mind-Body Problem* (Downers Grove, IL: InterVarsity Press, 2005), pp. 33–60.

Gould, Stephen Jay, 'Exaptation: A Crucial Tool for Evolutionary Analysis', *Journal of Social Issues* 47 (1991), pp. 43–65.

Gould, Stephen Jay, *Rocks of Ages: Science and Religion in the Fullness of Life* (London: Vintage, 2002).

Gould, Stephen J. and Richard C. Lewontin, 'The Spandrels of San Marco and the Panglossian Paradigm: A Critique of the Adaptationist Programme', *Proceedings of the Royal Society B* 205 (1979), pp. 581–98.

Graham, Keith, 'Altruism, Self-Interest and the Indistinctness of Persons', in Jonathan Seglow (ed.), *The Ethics of Altruism* (London: Frank Cass, 2004), pp. 49–67.

Granqvist, Pehr, et al., 'Sensed Presence and Mystical Experiences are Predicted by Suggestibility, Not by the Application of Transcranial Weak Complex Magnetic Fields', *Neuroscience Letters* 379.1 (2005), pp. 1–6.

Green, Joel B., 'Restoring the Human Person: New Testament Voices for a Wholistic and Social Anthropology', in Robert John Russell, et al. (eds.), *Neuroscience and the Person: Scientific Perspectives on Divine Action* (Vatican City: Vatican Observatory Publications/Berkeley, CA: Center for Theology and the Natural Sciences, 1999), pp. 3–22.

Green, Joel B. (ed.), *What about the Soul? Neuroscience and Christian Anthropology* (Nashville, TN: Abingdon, 2004).

Greene, Joshua D., 'The Secret Joke of Kant's Soul', in Walter Sinnott-Armstrong (ed.), *Moral Psychology, Vol. 3: The Neuroscience of Morality: Emotion, Brain Disorders, and Development* (Cambridge, MA: MIT Press, 2008), pp. 36–79.

Greene, Joshua D., 'Dual-Process Morality and the Personal/Impersonal Distinction: A Reply to McGuire, Langdon, Coltheart, and Mackenzie', *Journal of Experimental Social Psychology* 45 (2009), pp. 581–84.

Greene, Joshua D., 'Beyond Point-and-Shoot Morality: Why Cognitive (Neuro) Science Matters for Ethics', *Ethics* 124.4 (2014), pp. 695–726.

Greene, Joshua D., *Moral Tribes: Emotion, Reason, and the Gap between Us and Them* (London: Atlantic, 2014).

Greene, Joshua and Jonathan Cohen, 'For the Law, Neuroscience Changes Nothing and Everything', in Martha J. Farah (ed.), *Neuroethics: An Introduction with Readings* (Cambridge, MA: MIT Press, 2010), pp. 232–58.

Greene, Joshua and Jonathan Haidt, 'How (and Where) Does Moral Judgment Work?', *Trends in Cognitive Sciences* 6.12 (2002), pp. 517–23.

Greene, Joshua D., et al., 'An fMRI Investigation of Emotional Engagement in Moral Judgment', *Science* 293 (2001), pp. 2105–8.

Greene, Joshua D., et al., 'The Neural Bases of Cognitive Conflict and Control in Moral Judgment', *Neuron* 44 (2004), pp. 389–400.

Greene, Joshua D., et al., 'Cognitive Load Selectively Interferes with Utilitarian Moral Judgment', *Cognition* 107 (2008), pp. 1144–54.

Gunton, Colin, *The Triune Creator: A Historical and Systematic Study* (Grand Rapids, MI: Eerdmans, 1998).

Hagger, Martin S. and Nikos L. D. Chatzisarantis, 'A Multilab Preregistered Replication of the Ego-Depletion Effect', *Perspectives on Psychological Science* 11.4 (2016), pp. 546–73.

Haidt, Jonathan, 'The Emotional Dog and Its Rational Tail: A Social Intuitionist Approach to Moral Judgment', *Psychological Review* 108.4 (2001), pp. 814–34.

Haidt, Jonathan, 'The Emotional Dog Does Learn New Tricks: A Reply to Pizarro and Bloom (2003)', *Psychological Review* 110.1 (2003), pp. 197–98.

Haidt, Jonathan, *The Righteous Mind: Why Good People Are Divided by Politics and Religion* (London: Penguin, 2012).

Hallett, Mark, 'Volition: How Physiology Speaks to the Issue of Responsibility', in Walter Sinnott-Armstrong and Lynn Nadel (eds.), *Conscious Will and Responsibility: A Tribute to Benjamin Libet* (Oxford: Oxford University Press, 2010), pp. 61–69.

Hamilton, William D., 'The Genetical Evolution of Social Behaviour', parts 1 and 2, *Journal of Theoretical Biology* 7 (1964), pp. 1–52.

Hampshire, Adam, et al., 'Assessing Residual Reasoning Ability in Overtly Non-Communicative Patients Using fMRI', *Neuroimage: Clinical* 2 (2013), pp. 174–83.

Happel, Stephen, 'The Soul and Neuroscience: The Empirical Conditions for Human Agency and Divine Action', in Robert John Russell, et al. (eds.), *Neuroscience and the Person: Scientific Perspectives on Divine Action* (Vatican City: Vatican Observatory Publications/Berkeley, CA: Center for Theology and the Natural Sciences, 1999), pp. 281–304.

Hardin, Garrett, 'The Tragedy of the Commons', *Science* 168.3859 (1968), pp. 1243–48.

Harris, John, *Enhancing Evolution: The Ethical Case for Making Better People* (Princeton, NJ: Princeton University Press, 2007).

Harris, John, 'Is It Acceptable for People to Take Methylphenidate to Enhance Performance? Yes', *British Medical Journal* 338 (2009), pp. 1532–33.

Harris, John, 'Chemical Cognitive Enhancement: Is It Unfair, Unjust, Discriminatory, or Cheating for Healthy Adults to Use Smart Drugs?' in Judy Illes and Barbara J. Sahakian (eds.), *The Oxford Handbook of Neuroethics* (Oxford: Oxford University Press, 2011), pp. 265–72.

Hauerwas, Stanley, 'Must a Patient Be a Person to Be a Patient? Or, My Uncle Charlie Is Not Much of a Person but He Is Still My Uncle Charlie', *Connecticut Medicine* 39.12 (1975), pp. 815–17. Reprinted in *Journal of Religion, Disability and Health* 8.3–4 (2005), pp. 113–19.

Hauerwas, Stanley, with Richard Bondi and David B. Burrell, *Truthfulness and Tragedy: Further Investigations into Christian Ethics* (Notre Dame, IN: University of Notre Dame Press, 1977).

Hauskeller, Michael, 'Human Enhancement and the Giftedness of Life', *Philosophical Papers* 40.1 (2011), pp. 55–79.

Hauskeller, Michael, 'My Brain, My Mind, and I: Some Philosophical Assumptions of Mind-Uploading', *Journal of Machine Consciousness* 4.1 (2012), pp. 187–200.

Haynes, John-Dylan, 'Beyond Libet: Long-Term Prediction of Free Choices from Neuroimaging Signals', in Walter Sinnott-Armstrong and Lynn Nadel (eds.), *Conscious Will and Responsibility: A Tribute to Benjamin Libet* (Oxford: Oxford University Press, 2010), pp. 85–96.

Haynes, John-Dylan, 'Brain Reading: Decoding Mental States from Brain Activity in Humans', in Judy Illes and Barbara J. Sahakian (eds.), *The Oxford Handbook of Neuroethics* (Oxford: Oxford University Press, 2011), pp. 3–13.

Hays, Richard B., *The Moral Vision of the New Testament: Community, Cross and New Creation* (New York: HarperCollins, 1996).

Heap, Stephen, 'New Directions for Universities?', *Theology* 119.4 (2016), pp. 244–52.

Higton, Mike, *A Theology of Higher Education* (Oxford: Oxford University Press, 2012).

Hopkins, Patrick D., 'Why Uploading Will Not Work, or, the Ghosts Haunting Transhumanism', *Journal of Machine Consciousness* 4.1 (2012), pp. 229–44.

Horgan, Terry, 'The Phenomenology of Agency and the Libet Results', in Walter Sinnott-Armstrong and Lynn Nadel (eds.), *Conscious Will and Responsibility: A Tribute to Benjamin Libet* (Oxford: Oxford University Press, 2010), pp. 159–72.

Hughes, James, 'Contradictions from the Enlightenment Roots of Transhumanism', *Journal of Medicine and Philosophy* 35 (2010), pp. 622–40.

Hume, David, *A Treatise of Human Nature* (L. A. Selby-Bigge and P. H. Nidditch (eds.); Oxford: Clarendon, 1978).

Hurst, Samia, 'Clinical Research on Conditions Affecting Cognitive Capacity', in Judy Illes and Barbara J. Sahakian (eds.), *The Oxford Handbook of Neuroethics* (Oxford: Oxford University Press, 2011), pp. 513–28.

Huxley, T. H., 'Evolution and Ethics—Prolegomena', in *Evolution and Ethics and Other Essays* (Collected Essays, 9; London: Macmillan, 1894), pp. 1–45. Online at http://aleph0.clarku.edu/huxley/CE9/index.html

Hyman, Steven E., 'The Neurobiology of Addiction: Implications for Voluntary Control of Behavior', *American Journal of Bioethics* 7.1 (2007), pp. 8–11.

Hyman, Steven E., 'The Neurobiology of Addiction: Implications for Voluntary Control of Behavior', in Judy Illes and Barbara J. Sahakian (eds.), *The Oxford Handbook of Neuroethics* (Oxford: Oxford University Press, 2011), pp. 203–17.

Illes, Judy, and Barbara J. Sahakian (eds.), *The Oxford Handbook of Neuroethics* (Oxford: Oxford University Press, 2011).

Inzlicht, Michael and Elliot Berkman, 'Six Questions for the Resource Model of Control (and Some Answers)', *Social and Personality Psychology Compass* 9.10 (2015), pp. 511–24.

James, William, *The Varieties of Religious Experience: A Study in Human Nature* (New York: Longmans, Green and Co., 1917). Online at http://www.gutenberg.org/files/621/621-h/621-h.html

Jeeves, Malcolm (ed.), *From Cells to Souls – and Beyond: Changing Portraits of Human Nature* (Grand Rapids, MI: Eerdmans, 2004).

Jefferson, Will, et al., 'Enhancement and Civic Virtue', *Social Theory and Practice* 40.3 (2014), pp. 499–527.

Jennett, B. and F. Plum, 'Persistent Vegetative State after Brain Damage', *Lancet* 1 (1972), pp. 734–37.

Job, Veronika, Carol S. Dweck and Gregory M. Walton, 'Ego Depletion—Is It All in Your Head? Implicit Theories about Willpower Affect Self-Regulation', *Psychological Science* 21.11 (2010), pp. 1686–93.

John Paul II, Encyclical Letter *Evangelium Vitae* (25 March 1995). Online at http://w2.vatican.va/content/john-paul-ii/en/encyclicals/documents/hf_jp-ii_enc_25031995_evangelium-vitae.html

John Paul II, 'Address of John Paul II to the Participants in the International Congress on "Life-Sustaining Treatments and Vegetative State: Scientific Advances and Ethical Dilemmas"' (20 March 2004). Online at http://w2.vatican.va/content/john-paul-ii/en/speeches/2004/march/documents/hf_jp-ii_spe_20040320_congress-fiamc.html

Johnson, Dominic and Jesse Bering, 'Hand of God, Mind of Man: Punishment and Cognition in the Evolution of Co-operation', in Jeffrey Schloss and Michael J. Murray (eds.), *The Believing Primate: Scientific, Philosophical, and Theological' Reflections on the Origins of Religion* (Oxford: Oxford University Press, 2009), pp. 26–43.

Johnson, Dominic D. P., Hillary L. Lenfesty and Jeffrey P. Schloss, 'The Elephant in the Room: Do Evolutionary Accounts of Religion Entail the Falsity of Religious Belief?', *Philosophy, Theology and the Sciences* 1.2 (2014), pp. 200–31.

Jones, D. Gareth, 'The Emergence of Persons', in Malcolm Jeeves (ed.), *From Cells to Souls – and Beyond: Changing Portraits of Human Nature* (Grand Rapids, MI: Eerdmans, 2004), pp. 11–33.

Jonsen, Albert R., *The Birth of Bioethics* (Oxford: Oxford University Press, 1998).

Jotterand, Fabrice, 'Moral Enhancement, Neuroessentialism, and Moral Content', in Fabrice Jotterand and Veljko Dubljevic (eds.), *Cognitive Enhancement: Ethical and Policy Implications in International Perspectives* (Oxford: Oxford University Press, 2016), pp. 42–56.

Joyce, Richard, *The Evolution of Morality* (Cambridge, MA: MIT Press, 2006).

Juengst, Eric, 'What Does Enhancement Mean?' in Erik Parens (ed.), *Enhancing Human Traits: Ethical and Social Implications* (Washington, DC: Georgetown University Press, 1998), pp. 29–47.

Kahane, Guy, 'Evolutionary Debunking Arguments', *Noûs* 45.1 (2011a), pp. 103–25.

Kahane, Guy, 'Mastery without Mystery: Why there Is No Promethean Sin in Enhancement', *Journal of Applied Philosophy* 28.4 (2011b), pp. 355–68.

Kahane, Guy, 'On the Wrong Track: Process and Content in Moral Psychology', *Mind and Language* 27.5 (2012), pp. 519–45.

Kahane, Guy, 'Sidetracked by Trolleys: Why Sacrificial Moral Dilemmas Tell Us Little (or Nothing) About Utilitarian Judgment', *Social Neuroscience* 10.5 (2015), pp. 551–60.

Kahane, Guy and Julian Savulescu, 'Brain Damage and the Moral Significance of Consciousness', *Journal of Medicine and Philosophy* 34 (2009), pp. 6–26.

Kahane, Guy, et al., 'The Neural Basis of Intuitive and Counterintuitive Moral Judgment', *Social, Cognitive and Affective Neuroscience* 7 (2012), pp. 393–402.

Kamm, F. M., 'Neuroscience and Moral Reasoning: A Note on Recent Research', *Philosophy and Public Affairs* 37.4 (2009a), pp. 330–45.

Kamm, F. M., 'What Is and Is Not Wrong with Enhancement?', in Julian Savulescu and Nick Bostrom (eds.), *Human Enhancement* (Oxford: Oxford University Press, 2009b), pp. 91–130.

Kamm, F. M., with Judith Jarvis Thomson, Thomas Hurka, and Shelley Kagan, *The Trolley Problem Mysteries* (ed. Eric Rakowski; Oxford: Oxford University Press, 2015).

Kane, Robert, 'Some Neglected Pathways in the Free Will Labyrinth', in Robert Kane (ed.), *The Oxford Handbook of Free Will* (Oxford: Oxford University Press, 2002), pp. 407–37.

Kauppinen, Antti, 'Moral Sentimentalism', in Edward N. Zalta (ed.), *The Stanford Encyclopedia of Philosophy* (Spring 2014 Edition). Online at http://plato.stanford.edu/archives/spr2014/entries/moral-sentimentalism/

Kerr, Fergus, 'The Modern Philosophy of Self in Recent Theology', in Robert John Russell, et al. (eds.), *Neuroscience and the Person: Scientific Perspectives on Divine Action* (Vatican City: Vatican Observatory Publications/Berkeley, CA: Center for Theology and the Natural Sciences, 1999), pp. 23–40.

Kiesel, Andrea, et al., 'Unconscious Manipulation of Free Choice in Humans', *Consciousness and Cognition* 15 (2006), pp. 397–408.

Kilby, Karen, 'Perichoresis and Projection: Problems with Social Doctrines of the Trinity', *New Blackfriars* 81 (2000), pp. 432–45.

Kim, Jaegwon, *Mind in a Physical World* (Cambridge, MA: MIT Press, 1998).

Kim, Jaegwon, 'Blocking Causal Drainage and Other Maintenance Chores with Mental Causation', *Philosophy and Phenomenological Research* 67.1 (2003), pp. 151–76.

Koenigs, Michael, et al., 'Damage to the Prefrontal Cortex Increases Utilitarian Moral Judgements', *Nature* 446 (2007), pp. 908–11.

Koenigs, Michael, et al., 'Utilitarian Moral Judgment in Psychopathy', *Social, Cognitive and Affective Neuroscience* 7 (2012), pp. 708–14.

Kurzweil, Ray, *The Singularity Is Near* (London: Duckworth, 2005).

Larsson, Marcus, et al., 'Reply to M.A. Persinger and S. A. Koren's response to Granqvist et al. "Sensed Presence and Mystical Experiences are Predicted by Suggestibility, Not By the Application of Transcranial Weak Magnetic Fields"', *Neuroscience Letters* 380.3 (2005), pp. 348–50.

Laureys, Steven, 'The Neural Correlate of (Un)awareness: Lessons from the Vegetative State', *Trends in Cognitive Sciences* 9.12 (2005), pp. 556–59.

Laureys, Steven, et al., 'Auditory Processing in the Vegetative State', *Brain* 123.8 (2000), pp. 1589–601.

Laureys, Steven, et al., 'Cortical Processing of Noxious Somatosensory Stimuli in the Persistent Vegetative State', *Neuroimage* 17 (2002), pp. 732–41.

Laureys, Steven, et al., 'The Locked-In Syndrome: What Is It Like to be Conscious but Paralyzed and Voiceless?', *Progress in Brain Research* 150 (2005), pp. 495–511.

LeDoux, Joseph, et al. (eds.), *The Self: From Soul to Brain* (New York: New York Academy of Sciences, 2003).

Legrenzi, Paulo and Carlo Umtilà, *Neuromania: On the Limits of Brain Science* (Frances Anderson (trans.); Oxford: Oxford University Press, 2011).

Lenehan, Kevin A., '*Etsi Deus Non Daretur*: Bonhoeffer's Useful Misuse of Grotius' Maxim and Its Implications for Evangelisation in the World Come of Age', *The Bonhoeffer Legacy: Australasian Journal of Bonhoeffer Studies* 1.1 (2013), pp. 34–60.

Levy, Neil, 'Autonomy and Addiction', *Canadian Journal of Philosophy* 36 (2006a), pp. 427–48.

Levy, Neil, 'Addiction, Autonomy and Ego-Depletion: A Response to Bennett Foddy and Julian Savulescu', *Bioethics* 20.1 (2006b), pp. 16–20.

Levy, Neil, 'The Social: A Missing Term in the Debate over Addiction and Voluntary Control', *American Journal of Bioethics*, 7.1 (2007), pp. 35–36.

Levy, Neil, 'Going beyond the Evidence', *American Journal of Bioethics* 8.9 (2008), pp. 19–21.

Levy, Neil and Julian Savulescu, 'Moral Significance of Phenomenal Consciousness', *Progress in Brain Research* 177 (2009), pp. 361–70.

Levy, Neil, et al., 'Are You Morally Modified? The Moral Effects of Widely Used Pharmaceuticals', *Philosophy, Psychiatry, and Psychology* 21.2 (2014), pp. 111–25.

Libet, Benjamin, 'Do We Have Free Will?' *Journal of Consciousness Studies* 6 (1999), pp. 47–57.

Locke, John, *An Essay Concerning Human Understanding* (Peter H. Nidditch (ed.); The Clarendon Edition of the Works of John Locke; Oxford: Clarendon, 1975).

Logothetis, Nikos K., 'What We Can Do and What We Cannot Do with fMRI', *Nature* 453 (2008), pp. 869–78.

Luther, Martin, 'Disputation against Scholastic Theology', in *Luther: Early Theological Works* (James Atkinson (trans.); Library of Christian Classics, 16; London: SCM, 1962), pp. 266–73.

Lysaught, M. Therese and Joseph J. Kotva, Jr., with Stephen E. Lammers and Allen Verhey (eds.), *On Moral Medicine: Theological Perspectives in Medical Ethics* (Grand Rapids, MI: Eerdmans, 3rd edn, 2012).

McCormack, Bruce L., *Orthodox and Modern: Studies in the Theology of Karl Barth* (Grand Rapids, MI: Baker Academic, 2008).

McFadyen, Alistair I., *The Call to Personhood: A Christian Theory of the Individual in Social Relationships* (Cambridge: Cambridge University Press, 1990).

McFadyen, Alistair I., *Bound to Sin: Abuse, Holocaust and the Christian Doctrine of Sin* (Cambridge: Cambridge University Press, 2000).

McFarland, Ian, 'Who Is My Neighbor? The Good Samaritan as a Source for Theological Anthropology', *Modern Theology* 17.1 (2001), pp. 57–66.

McFarland, Ian, '"Willing Is Not Choosing": Some Anthropological Implications of Dyothelite Christology', *International Journal of Systematic Theology* 9.1 (2007), pp. 3–23.

McGuire, Jonathan, et al., 'A Reanalysis of the Personal/Impersonal Distinction in Moral Psychology Research', *Journal of Experimental Social Psychology* 45 (2009), pp. 577–80.

MacIntyre, Alasdair, *After Virtue: A Study in Moral Theory* (London: Duckworth, 2nd edn, 1985).

McIntyre, Alison, 'Doctrine of Double Effect', in Edward N. Zalta (ed.), *The Stanford Encyclopedia of Philosophy* (Winter 2014 Edition). Online at http://plato.stanford.edu/archives/win2014/entries/double-effect

McKenny, Gerald P., *To Relieve the Human Condition: Bioethics, Technology, and the Body* (Albany: State University of New York Press, 1997).

McNamara, Patrick, *The Neuroscience of Religious Experience* (Cambridge: Cambridge University Press, 2009).

Maher, Brendan, 'Poll Results: Look Who's Doping', *Nature* 452.7188 (2008), pp. 674–75.

Martin, Amanda, et al., 'A Four-Part Working Bibliography of Neuroethics: Part 3 – "second tradition neuroethics" – Ethical Issues in Neuroscience', *Philosophy, Ethics, and Humanities in Medicine* 11.7 (2016), DOI: 10.1186/s13010-016-0037-1.

Matsuhashi, Masao and Mark Hallett, 'The Timing of the Conscious Intention to Move', *European Journal of Neuroscience* 28 (2008), pp. 2344–51.

Matthews, Pia, *Pope John Paul II and the Apparently 'Non-Acting' Person* (Leominster: Gracewing, 2013).

Megone, Christopher, 'Aristotle's Function Argument and the Concept of Mental Illness', *Philosophy, Psychiatry, and Psychology* 5.3 (1998), pp. 187–201.

Megone, Christopher, 'Mental Illness, Human Function, and Values', *Philosophy, Psychiatry, and Psychology* 7.1 (2000), pp. 45–65.

Messer, Neil, *Selfish Genes and Christian Ethics: Theological and Ethical Reflections on Evolutionary Biology* (London: SCM Press, 2007).

Messer, Neil, *Respecting Life: Theology and Bioethics* (London: SCM, 2011).

Messer, Neil, *Flourishing: Health, Disease and Bioethics in Theological Perspective* (Grand Rapids, MI: Eerdmans, 2013).

Metzinger, Thomas and Elisabeth Hildt, 'Cognitive Enhancement', in Judy Illes and Barbara J. Sahakian (eds.), *The Oxford Handbook of Neuroethics* (Oxford: Oxford University Press, 2011), pp. 245–64.

Midgley, Mary, 'Why Memes?' in Hilary Rose and Steven Rose (eds.), *Alas Poor Darwin: Arguments against Evolutionary Psychology* (London: Jonathan Cape, 2000), pp. 67–84.

Mill, John Stuart, *On Liberty* (1863), in *Utilitarianism, On Liberty, Considerations on Representative Government, Remarks on Bentham's Philosophy* (Geraint Williams (ed.); London: J. M. Dent, 3rd edn, 1993).

Monti, Martin M., et al., 'Willful Modulation of Brain Activity in Disorders of Consciousness', *New England Journal of Medicine* 362.7 (2010), pp. 579–89.

Moore, Michael S., 'Libet's Challenge(s) to Responsible Agency', in Walter Sinnott-Armstrong and Lynn Nadel (eds.), *Conscious Will and Responsibility: A Tribute to Benjamin Libet* (Oxford: Oxford University Press, 2010), pp. 207–30.

Morse, Stephen J., 'Criminal Law and Common Sense: An Essay on the Perils and Promise of Neuroscience', *Marquette Law Review* 99.1 (2015), pp. 39–74.

Moya, Carlos J., 'Mind, Brain, and Downward Causation', in Juan José Sanguineti, Aribierto Acerbi and José Angel Lombo (eds.), *Moral Behavior and Free Will: A Neurobiological and Philosophical Approach* (Morolo: If Press, 2011), pp. 185–200.

Mühling, Markus, *Resonances: Neurobiology, Evolution and Theology. Evolutionary Niche Construction, the Ecological Brain and Relational-Narrative Theology* (Göttingen: Vandenhoeck and Ruprecht, 2014).

Multi-Society Task Force on the Persistent Vegetative State, 'Medical Aspects of a Persistent Vegetative State', *New England Journal of Medicine* 330 (1994), pp. 499–508, 572–79.

Murphy, Nancey, *Bodies and Souls, or Spirited Bodies* (Cambridge: Cambridge University Press, 2006).

Murphy, Nancey, 'Avoiding Neurobiological Reductionism: The Role of Downward Causation in Complex Systems', in Juan José Sanguineti, Aribierto Acerbi and José Angel Lombo (eds.), *Moral Behavior and Free Will: A Neurobiological and Philosophical Approach* (Morolo: If Press, 2011), pp. 201–22.

Murphy, Nancey, 'Do Humans Have Souls? Perspectives from Philosophy, Science, and Religion', *Interpretation* 67.1 (2013), pp. 30–41.

Murphy, Nancey and Warren S. Brown, *Did My Neurons Make Me Do It? Philosophical and Neurobiological Perspectives on Moral Responsibility and Free Will* (Oxford: Oxford University Press, 2007).

Murray, Michael J., 'Scientific Explanations of Religion and the Justification of Religious Belief', in Jeffrey Schloss and Michael J. Murray (eds.), *The Believing Primate: Scientific, Philosophical, and Theological Reflections on the Origins of Religion* (Oxford: Oxford University Press, 2009), pp. 168–78.

Murray, Michael J. and Andrew Goldberg, 'Evolutionary Accounts of Religion: Explaining and Explaining Away', in Jeffrey Schloss and Michael J. Murray (eds.), *The Believing Primate: Scientific, Philosophical, and Theological Reflections on the Origins of Religion* (Oxford: Oxford University Press, 2009), pp. 179–99.

Nadelhoffer, Thomas, 'The Threat of Shrinking Agency and Free Will Disillusionism', in Walter Sinnott-Armstrong and Lynn Nadel (eds.), *Conscious Will and Responsibility: A Tribute to Benjamin Libet* (Oxford: Oxford University Press, 2010), pp. 173–88.

Nagel, Thomas, 'What Is It Like to Be a Bat?', *Philosophical Review* 83.4 (1974), pp. 435–50.

National Commission for the Protection of Human Subjects, *The Belmont Report: Ethical Principles and Guidelines for the Protection of Human Subjects of Research* (DHEW Publication No. (OS) 78-0012; Washington, DC: U.S. Government Printing Office, 1978).

Newberg, Andrew B., *Principles of Neurotheology* (Farnham: Ashgate, 2010).

Newberg, Andrew B., et al., 'The Measurement of Regional Cerebral Blood Flow During the Complex Cognitive Task of Meditation: A Preliminary SPECT Study', *Psychiatry Research: Neuroimaging* 106 (2001), pp. 113–22.

Newberg, Andrew B., et al., 'Cerebral Blood Flow during Meditative Prayer: Preliminary Findings and Methodological Issues', *Perceptual and Motor Skills* 97.2 (2003), pp 625–30.

Newberg, Andrew B., et al., 'The Measurement of Regional Cerebral Blood Flow During Glossolalia: A Preliminary SPECT Study', *Psychiatry Research: Neuroimaging* 148.1 (2006), pp. 67–71.

Newberg, Andrew B., et al., 'Cerebral Blood Flow Differences between Long-Term Meditators and Non-Meditators', *Consciousness and Cognition* 19.4 (2010), pp. 899–905.

Nordenfelt, Lennart, *On the Nature of Health: An Action-Theoretic Approach* (Dordrecht: Kluwer Academic, 2nd edn, 1995).

Nuffield Council on Bioethics, *Novel Neurotechnologies: Intervening in the Brain* (London: Nuffield Council on Bioethics, 2013).

O'Donovan, Oliver, 'Again: Who Is a Person?', in M. Therese Lysaught and Joseph J. Kotva, Jr., with Stephen E. Lammers and Allen Verhey (eds.), *On Moral Medicine: Theological Perspectives in Medical Ethics* (Grand Rapids, MI: Eerdmans, 3rd edn, 2012), pp. 367–71.

O'Donovan, Oliver, *Self, World, and Time: Ethics as Theology* (Grand Rapids, MI: Eerdmans, 2013).

O'Neill, Onora, *Autonomy and Trust in Bioethics* (Cambridge: Cambridge University Press, 2002).

Oliver, Simon, 'Aquinas and Aristotle's Teleology', *Nova et Vetera* 11.3 (2013), pp. 849–70.

Owen, Adrian M. and Martin R. Coleman, 'Functional Neuroimaging of the Vegetative State', *Nature Reviews Neuroscience* 9 (2008), pp. 235–43.

Owen, Adrian M., et al., 'Residual Auditory Function in Persistent Vegetative State: A Combined PET and fMRI Study', *Neuropsychological Rehabilitation* 15.3/4 (2005), pp. 290–306.

Owen, Adrian M., et al., 'Detecting Awareness in the Vegetative State', *Science* 313.5792 (2006), p. 1402.

Parens, Erik, 'Toward a More Fruitful Debate about Enhancement', in Julian Savulescu and Nick Bostrom (eds.), *Human Enhancement* (Oxford: Oxford University Press, 2009), pp. 181–97.

Paxton, Joseph M., Tommaso Bruni and Joshua D. Greene, 'Are "counter-intuitive" Deontological Judgments Really Counter-Intuitive? An Empirical Reply to Kahane et al. (2012)', *Social, Cognitive and Affective Neuroscience* 9 (2014), pp. 1368–71.

Peacocke, Arthur, 'Science and the Future of Theology: Critical Issues', *Zygon* 35.1 (2000), pp. 119–40.

Perry, John (ed.), *A Time to Heal: A Report for the House of Bishops on the Healing Ministry* (London: Church House Publishing, 2000).

Perry, John, 'Jesus and Hume among the Neuroscientists: Haidt, Greene, and the Unwitting Return of Moral Sense Theory', *Journal of the Society of Christian Ethics* 36.1 (2016), pp. 69–85.

Persinger, Michael A., 'Religious and Mystical Experiences as Artifacts of Temporal Lobe Function: A General Hypothesis', *Perceptual and Motor Skills* 57 (1983), pp. 1255–62.

Persinger, M. A. and S. A. Koren, 'A Response to Granqvist et al. "Sensed Presence and Mystical Experiences are Predicted by Suggestibility, Not By the Application of Transcranial Weak Magnetic Fields"', *Neuroscience Letters* 380.3 (2005), pp. 346–47.

Persson, Ingmar and Julian Savulescu, 'The Perils of Cognitive Enhancement and the Urgent Imperative to Enhance the Moral Character of Humanity', *Journal of Applied Philosophy* 25.3 (2008), pp. 162–77.

Peters, Ted, 'Resurrection of the Very Embodied Soul?', in Robert John Russell et al. (eds.), *Neuroscience and the Person: Scientific Perspectives on Divine Action* (Vatican City: Vatican Observatory Publications/Berkeley, CA: Center for Theology and the Natural Sciences, 1999), pp. 305–26.

Peterson, Gregory R., *Minding God: Theology and the Cognitive Sciences* (Minneapolis, MN: Fortress, 2003).

Peterson, Gregory R., 'Are Evolutionary/Cognitive Theories of Religion Relevant for Philosophy of Religion?' *Zygon* 45.3 (2010), pp. 545–57.

Pizzaro, David A. and Paul Bloom, 'The Intelligence of the Moral Intuitions: Comment on Haidt (2001)', *Psychological Review* 110.1 (2003), pp. 193–96.

Plantinga, Alvin, 'Games Scientists Play', in Jeffrey Schloss and Michael J. Murray (eds.), *The Believing Primate: Scientific, Philosophical, and Theological Reflections on the Origins of Religion* (Oxford: Oxford University Press, 2009), pp. 139–67.

Pockett, Susan and Suzanne C. Purdy, 'Are Voluntary Movements Initiated Preconsciously? The Relationships between Readiness Potentials, Urges, and Decisions', in Walter Sinnott-Armstrong and Lynn Nadel (eds.), *Conscious Will and Responsibility: A Tribute to Benjamin Libet* (Oxford: Oxford University Press, 2010), pp. 34–46.

President's Council on Bioethics, Transcript of Session 2, 'Science and the Pursuit of Perfection', 17 January 2002. Online at https://bioethicsarchive.georgetown.edu/pcbe/transcripts/jan02/jan17session2.html

President's Council on Bioethics, Transcript of Session 2, 'Choosing the Sex of Children: Demographics', 17 October 2002. Online at https://bioethicsarchive.georgetown.edu/pcbe/transcripts/oct02/session2.html

President's Council on Bioethics, *Beyond Therapy: Biotechnology and the Pursuit of Happiness* (Washington, DC: President's Council on Bioethics, 2003). Online at https://repository.library.georgetown.edu/handle/10822/559341

Pritchard, Kathryn, 'Bioethics, Public Policy and the Church of England' (unpublished doctoral dissertation, University of Winchester, 2015).

Racine, Eric and Emily Bell, 'Clinical and Public Translation of Neuroimaging Research in Disorders of Consciousness Challenges Current Diagnostic and Public Understanding Paradigms', *American Journal of Bioethics* 8.9 (2008), pp. 13–15.

Ragan, C. Ian, Imre Bard and Ilina Singh, 'What Should We Do about Student Use of Cognitive Enhancers? An Analysis of Current Evidence', *Neuropharmacology* 64 (2013), pp. 588–95.

Rawls, John, *A Theory of Justice* (Cambridge, MA: Harvard University Press, 1971).

Reed, Esther D., *The Ethics of Human Rights: Contested Doctrinal and Moral Issues* (Waco, TX: Baylor University Press, 2007).

Reed, Esther D., 'Responsibility to Protect and Militarized Intervention: When and Why the Churches Failed to Discern Moral Hazard', *Journal of Religious Ethics* 40.2 (2012), pp. 308–34.

Reinders, Hans S., *Receiving the Gift of Friendship: Profound Disability, Theological Anthropology, and Ethics* (Grand Rapids, MI: Eerdmans, 2008).

Reinders, Hans S., 'Understanding Humanity and Disability: Probing an Ecological Perspective', *Studies in Christian Ethics* 26.1 (2013), pp. 37–49.

Reynolds, Thomas E., *Vulnerable Communion: A Theology of Disability and Hospitality* (Grand Rapids, MI: Brazos, 2008).

Rose, Hilary and Steven Rose (eds.), *Alas Poor Darwin: Arguments against Evolutionary Psychology* (London: Jonathan Cape, 2000).

Roser, Matthew and Michael S. Gazzaniga, 'Automatic Brains – Interpretive Minds', *Current Directions in Psychological Science* 13.2 (2004), pp. 56–59.

Roskies, Adina, 'Neuroethics for the New Millennium', *Neuron* 35 (2002), pp. 21–23.

Roskies, Adina, 'Neuroimaging and Inferential Distance', *Neuroethics* 1.1 (2008), pp. 19–30.

Roskies, Adina, 'Why Libet's Studies Don't Pose a Threat to Free Will', in Walter Sinnott-Armstrong and Lynn Nadel (eds.), *Conscious Will and Responsibility: A Tribute to Benjamin Libet* (Oxford: Oxford University Press, 2010), pp. 11–22.

Rughani, Anand E. and Paul L. Penar, 'Brain Anatomy'. Online at http://emedicine.medscape.com/article/1898830-overview

Russell, Robert John, et al. (eds.), *Neuroscience and the Person: Scientific Perspectives on Divine Action* (Vatican City: Vatican Observatory Publications/Berkeley, CA: Center for Theology and the Natural Sciences, 1999).

Sahakian, Barbara and Sharon Morein-Zamir, 'Professor's Little Helper', *Nature* 450 (2007), pp. 1157–59.

St-Pierre, L. S. and M. A. Persinger, 'Experimental Facilitation of the Sensed Presence Is Predicted by the Specific Patterns of the Applied Magnetic Fields, Not by Suggestibility: Re-Analyses of 19 Experiments', *International Journal of Neuroscience* 116 (2006), pp. 1079–96.

Sandberg, Anders, 'Cognition Enhancement: Upgrading the Brain', in Julian Savulescu, Ruud ter Meulen and Guy Kahane (eds.), *Enhancing Human Capacities* (Chichester: Wiley-Blackwell, 2011), pp. 71–91.

Sandberg, Anders and Julian Savulescu, 'The Social and Economic Impacts of Cognitive Enhancement', in Julian Savulescu, Ruud ter Meulen and Guy Kahane (eds.), *Enhancing Human Capacities* (Chichester: Wiley-Blackwell, 2011), pp. 92–112.

Sandel, Michael J., 'The Case against Perfection: What's Wrong with Designer Children, Bionic Athletes, and Genetic Engineering', *Atlantic Monthly* 292.3 (2004). Reprinted in Julian Savulescu and Nick Bostrom (eds.), *Human Enhancement* (Oxford: Oxford University Press, 2009), pp. 71–89.

Sandel, Michael J., *The Case against Perfection: Ethics in the Age of Genetic Engineering* (Cambridge, MA: Belknap Press, 2007).

Sarot, Marcel, 'Christian Faith, Free Will and Neuroscience', in Peter Jonkers and Marcel Sarot (ed.), *Embodied Religion: Proceedings of the 2012 Conference of the European Society for Philosophy of Religion* (Utrecht: Igitur, 2013), pp. 105–20.

Schiff, Nicholas, et al., 'Words without Mind', *Journal of Cognitive Neuroscience* 11.6 (1999), pp. 650–56.

Schiff, Nicholas, et al., 'fMRI Reveals Large-Scale Network Activation in Minimally Conscious Patients', *Neurology* 64.3 (2005), pp. 514–23.

Schiff, Nicholas, et al., 'Behavioural Improvements with Thalamic Stimulation after Severe Traumatic Brain Injury', *Nature* 448 (2007), pp. 600–3.

Schjoedt, Uffe, 'The Religious Brain: A General Introduction to the Experimental Neuroscience of Religion', *Method and Theory in the Study of Religion* 21 (2009), pp. 310–39.

Schjoedt, Uffe, et al., 'The Power of Charisma – Perceived Charisma Inhibits the Frontal Executive Network of Believers in Intercessory Prayer', *Social, Cognitive and Affective Neurosciences* 6.1 (2011), pp. 119–27.

Schloss, Jeffrey, 'Introduction: Evolutionary Theories of Religion. Science Unfettered or Naturalism Run Wild?' in Jeffrey Schloss and Michael J. Murray (eds.), *The Believing Primate: Scientific, Philosophical, and Theological Reflections on the Origins of Religion* (Oxford: Oxford University Press, 2009), pp. 1–25.

Schwartz, Rob and Mirra Schwartz, 'The Risks of Reducing Consciousness to Neuroimaging', *American Journal of Bioethics* 8.9 (2008), pp. 25–26.

Severino, Sally K., *Behold Our Moral Body: Psychiatry, Duns Scotus, and Neuroscience* (London: Versita, 2013).

Shafer, Michael R., *Well Played: A Christian Theology of Sport and the Ethics of Doping* (Eugene, OR: Pickwick, 2015).

Shaw, P., et al., 'Intellectual Ability and Cortical Development in Children and Adolescents', *Nature* 440.7084 (2006), pp. 676–79.

Simons, Daniel J. and Christopher F. Chabris, 'Gorillas in our Midst: Sustained Inattentional Blindness for Dynamic Events', *Perception* 28.9 (1999), pp. 1059–74.

Simpson, Peter, 'The Definition of Person: Boethius Revisited', *The New Scholasticism* 62.2 (1988), pp. 210–20.

Singer, Peter, 'Ethics and Intuitions', *Journal of Ethics* 9 (2005), pp. 331–52.

Singh, Ilina, Imre Bard and Jonathan Jackson, 'Robust Resilience and Substantial Interest: A Survey of Pharmacological Cognitive Enhancement among University Students in the UK and Ireland', *PLoS ONE* 9.10 (2014), doi: 10.1371/journal.pone.0105969.

Sinnott-Armstrong, Walter and Lynn Nadel, 'Introduction', in Walter Sinnott-Armstrong and Lynn Nadel (eds.), *Conscious Will and Responsibility: A Tribute to Benjamin Libet* (Oxford: Oxford University Press, 2010), pp. xi–xv.

Smith, Christian, *Moral, Believing Animals: Human Personhood and Culture* (Oxford: Oxford University Press, 2003).

Smith, M. Elizabeth and Martha J. Farah, 'Are Prescription Stimulants "Smart Pills"? The Epidemiology and Cognitive Neuroscience of Prescription Stimulant Use by Normal Healthy Individuals', *Psychological Bulletin* 137.5 (2011), pp. 717–41.

Sober, Elliot and David Sloan Wilson, *Unto Others: The Evolution and Psychology of Unselfish Behavior* (Cambridge, MA: Harvard University Press, 1998).

Sorger, Bettina, et al., 'A Real-Time fMRI-Based Spelling Device Immediately Enabling Robust Motor-Independent Communication', *Current Biology* 22 (2012), pp. 1333–38.

Spezio, Michael L., 'The Neuroscience of Emotion and Reasoning in Social Contexts: Implications for Moral Theology', *Modern Theology* 27.2 (2011), pp. 339–56.

Strawson, Galen, 'The Bounds of Freedom', in Robert Kane (ed.), *The Oxford Handbook of Free Will* (Oxford: Oxford University Press, 2002), pp. 441–60.

Street, Sharon, 'A Darwinian Dilemma for Realist Theories of Value', *Philosophical Studies* 127 (2006), pp. 109–66.

Swanson, James M., et al., 'Attention Deficit Hyperactivity Disorder: Defining a Spectrum Disorder and Considering Neuroethical Implications', in Judy Illes and Barbara J. Sahakian (eds.), *The Oxford Handbook of Neuroethics* (Oxford: Oxford University Press, 2011), pp. 309–40.

Swinton, John, 'Known by God', in Hans Reinders (ed.), *The Paradox of Disability: Responses to Jean Vanier and the L'Arche Community from Theology and the Sciences* (Grand Rapids, MI: Eerdmans, 2010).

Swinton, John, *Dementia: Living in the Memories of God* (Grand Rapids, MI: Eerdmans, 2012).

Taylor, Charles, *Sources of the Self: The Making of the Modern Identity* (Cambridge: Cambridge University Press, 1989).

Thomson, Judith Jarvis, 'The Trolley Problem', *Yale Law Journal* 94.6 (1985), pp. 1395–415.

Timmons, Mark, 'Toward a Sentimentalist Deontology', in Walter Sinnott-Armstrong (ed.), *Moral Psychology, Vol. 3: The Neuroscience of Morality: Emotion, Brain Disorders, and Development* (Cambridge, MA: MIT Press, 2008), pp. 93–104.

Tononi, Giulio, 'Consciousness as Integrated Information: A Provisional Manifesto', *Biological Bulletin* 215 (2008), pp. 216–42.

Torrance, Alan, 'What Is a Person?', in Malcolm Jeeves (ed.), *From Cells to Souls – and Beyond: Changing Portraits of Human Nature* (Grand Rapids, MI: Eerdmans, 2004a), pp. 199–222.

Torrance, Alan, 'Developments in Neuroscience and Human Freedom: Some Theological and Philosophical Questions', *Science and Christian Belief* 16.2 (2004b), pp. 123–37.

Trivers, Robert L., 'The Evolution of Reciprocal Altruism', *Quarterly Review of Biology* 46 (1971), pp. 35–56.

van Gulick, Robert, 'Consciousness', in Edward N. Zalta (ed.), *The Stanford Encyclopedia of Philosophy* (Spring 2014 Edition). Online at http://plato.stanford.edu/archives/spr2014/entries/consciousness/

van Inwagen, Peter, 'Explaining Belief in the Supernatural: Some Thoughts on Paul Bloom's "Religious Belief as an Evolutionary Accident"', in Jeffrey Schloss and Michael J. Murray (eds.), *The Believing Primate: Scientific, Philosophical, and*

Theological Reflections on the Origins of Religion (Oxford: Oxford University Press, 2009), pp. 128–38.

Van Slyke, James A., et al. (eds.), *Theology and the Science of Moral Action: Virtue Ethics, Exemplarity, and Cognitive Neuroscience* (New York: Routledge, 2013).

Visala, Aku, 'Theism, Compatibilism and Neurodeterminism: A Response to Marcel Sarot', in Peter Jonkers and Marcel Sarot (eds.), *Embodied Religion: Proceedings of the 2012 Conference of the European Society for Philosophy of Religion* (Utrecht: Igitur, 2013), pp. 121–39.

Volkow, Nora D., et al., 'A Decade after The Decade of the Brain', *Cerebrum* (26 February 2010). Online at http://dana.org/Cerebrum/2010/A_Decade_after_The_Decade_of_the_Brain__Compilation/

Wakefield, Jerome C., 'The Concept of Mental Disorder: On the Boundary between Biological Facts and Social Values', *American Psychologist* 47.3 (1992), pp. 373–88.

Walter, Sven, 'Evolutionary Psychology', in James Fieser and Bradley Dowden (eds.), *Internet Encyclopedia of Philosophy* (n.d.). Online at http://www.iep.utm.edu/evol-psy/

Wang, Danny J. J., et al., 'Cerebral Blood Flow Changes Associated with Different Meditation Practices and Perceived Depth of Meditation', *Psychiatry Research: Neuroimaging* 191 (2011), pp. 60–67.

Ward, Keith, *In Defence of the Soul* (Oxford: Oneworld, 1998).

Warren, Mary Anne, 'On the Moral and Legal Status of Abortion', *The Monist* 57.1 (1973), pp. 43–61.

Waters, Brent, *From Human to Posthuman: Christian Theology and Technology in a Postmodern World* (Farnham: Ashgate, 2006).

Waters, Brent, 'Christian Ethics and Human Germline Genetic Modification', *Christian Bioethics* 18.2 (2012), pp. 115–25.

Waters, Brent, *Christian Moral Theology in the Emerging Technoculture: From Posthuman Back to Human* (Farnham: Ashgate, 2014).

Wegner, Daniel M., *The Illusion of Conscious Will* (Cambridge, MA: MIT Press, 2002).

White, Roger M., '"Ought" Implies "Can": Kant and Luther, a Contrast', in George MacDonald Ross and Tony McWalter (eds.), *Kant and His Influence* (Bristol: Thoemmes, 1990), pp. 1–72.

Wilkinson, John, *The Bible and Healing: A Medical and Theological Commentary* (Grand Rapids, MI: Eerdmans, 1998).

Wilkinson, Dominic, Guy Kahane and Julian Savulescu, '"Neglected Personhood" and Neglected Questions: Remarks on the Moral Significance of Consciousness', *American Journal of Bioethics* 8.9 (2008), pp. 31–33.

Wilkinson, D. J., et al., 'Functional Neuroimaging and Withdrawal of Life-Sustaining Treatment from Vegetative Patients', *Journal of Medical Ethics* 35 (2009), pp. 508–11.

Williams, George C., *Adaptation and Natural Selection: A Critique of Some Current Evolutionary Thought* (Princeton, NJ: Princeton University Press, 1966).

Williams, Rowan, 'What Is a University?' (lecture given at Wuhan University, China, 13 October 2006). Online at http://rowanwilliams.archbishopofcanterbury.org/articles.php/1468/china-universities-have-essential-role-in-public-life

Wilson, David Sloan, *Darwin's Cathedral: Evolution, Religion, and the Nature of Society* (Chicago, IL: University of Chicago Press, 2002).

Wilson, David Sloan, 'Evolutionary Social Constructivism: Narrowing (but Not Yet Bridging) the Gap', in Jeffrey Schloss and Michael J. Murray (eds.), *The Believing Primate: Scientific, Philosophical, and Theological Reflections on the Origins of Religion* (Oxford: Oxford University Press, 2009), pp. 319–38.

Wiseman, Harris, *The Myth of the Moral Brain: The Limits of Moral Enhancement* (Cambridge, MA: MIT Press, 2016).

Wolpe, Paul Root, 'Is My Mind Mine? Neuroethics and Brain Imaging', in Vardit Ravitsky, Autumn Fiester and Arthur L. Caplan (eds.), *The Penn Center Guide to Bioethics* (New York, NY: Springer, 2009). Online at http://search.credoreference.com/content/entry/sppbioeth/is_my_mind_mine_neuroethics_and_brain_imaging/0

World Health Organization Expert Committee on Addiction-Producing Drugs, *Thirteenth Report* (World Health Organization Technical Report Series, 273: Geneva: World Health Organization, 1964). Online at http://www.who.int/medicines/areas/quality_safety/reports/en/

World Health Organization, *Basic Documents* (Geneva: World Health Organization, 48th edn, 2014). Online at http://apps.who.int/gb/bd/

Wren, Brian, 'Great God, Your Love Has Called Us Here', in *Rejoice and Sing* (Oxford: Oxford University Press, 1991), no. 339.

Wright, Robert, *The Moral Animal: Evolutionary Psychology and Everyday Life* (London: Abacus, 1996).

Wu, Kevin Chien-Chang, 'Soul-Making in Neuroimaging?' *American Journal of Bioethics* 8.9 (2008), pp. 21–22.

Zagzebski, Linda, 'The Uniqueness of Persons', *Journal of Religious Ethics* 29.3 (2001), pp. 401–23.

Index

CPSIA information can be obtained
at www.ICGtesting.com
Printed in the USA
LVOW13*1916090518
576575LV00009B/117/P